# The ABC of VIDEO MOVIES

Getting the best from your camcorder

AF677914

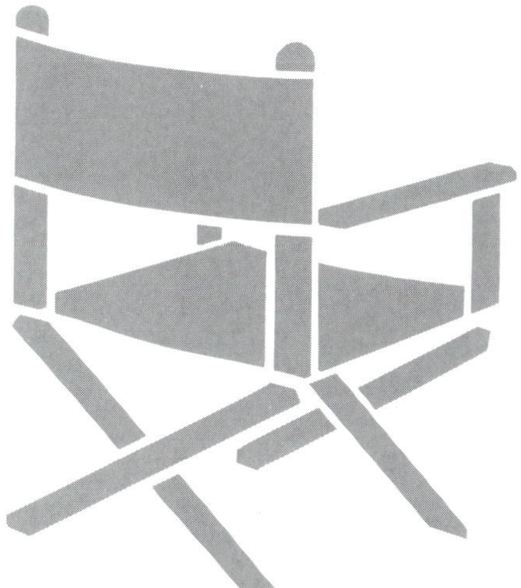

# The ABC of VIDEO MOVIES

**Getting the best from your camcorder**

Alan Cleave

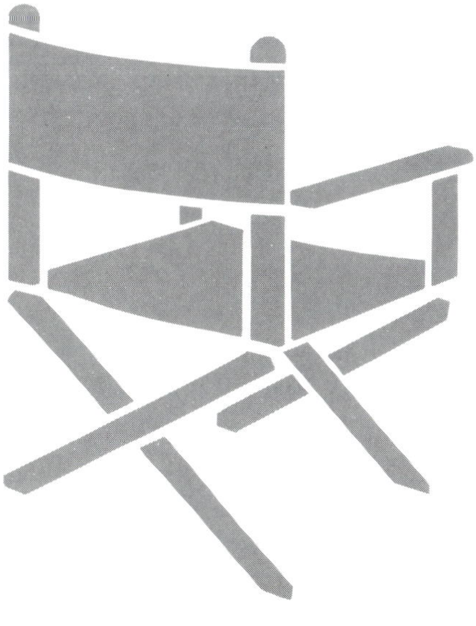

R. Morgan Publishing

©1988 Alan Cleave
All Rights Reserved

Published by:

R. Morgan Publishing
PO Box 11
Chislehurst
Kent BR7 5RH

British Library Cataloguing in Publication Data

Cleave, Alan
The ABC of video movies: getting the best from your camcorder.
1. Videorecording — Manuals
I. Title
778.59'9
ISBN 0-902979-06-X

Printed in the United Kingdom by

Manor Park Press Ltd
Highfield Industrial Estate,
Eastbourne, East Sussex

# CONTENTS

# PREFACE

This book is aimed at those who have already purchased video equipment and want to get the most satisfaction from it. So it contains no advice on choosing and buying a camcorder (there are some excellent magazines that can give up-to-the-minute guidance on that topic); nor does it tell you how to operate individual pieces of equipment (the instructions that come with it will explain all that).

But what it *will* do is show how you can use your apparatus to make really satisfying video movies that you can be proud of and which your viewers will enjoy watching.

It's a mistaken belief that to make a professional-looking movie you need professional equipment. Nothing could be further from the truth. With a simple camcorder, it's possible to produce results that, for most subjects you're likely to tackle, can hardly be distinguished from those achieved by the 'big boys'. The secret of success lies in understanding the relatively simple rules that govern the motion picture medium and applying them with imagination.

That is what I hope this book will help you to do.

Alan Cleave.

# Chapter 1
# The Language of Pictures

## What is a movie?

That may seem a pretty silly question. After all, aren't we seeing movies every day on a television screen or at the cinema? Of course we are — and perhaps that's really the trouble. The motion picture experience is now such a common one that we accept it without consciously querying its essential nature.

And of course there's no reason why most people should — if movies, for them, are just an agreeable entertainment or a means of gaining information. They don't need to understand *how* a movie communicates, anymore than they need to know how their quartz wrist-watch functions. But this passive, unquestioning attitude is hardly appropriate for those who aspire to make movies themselves, no matter how modest their ambitions may be. We don't come into this world with an innate understanding of how a video should be put together; this is a craft that has to be learnt just like making a chair or baking a cake.

# 1

You think I'm labouring the obvious in saying this? Not a bit. There are people everywhere who think they know all about making movies without ever having had five minutes formal instruction on the subject. Nobody in their right senses would attempt to write a string quartet without at least a modicum of musical training, yet in recent years tens of thousands have purchased video cameras and sallied forth believing that the mere acquisition of equipment has turned them into fully-fledged movie makers. Of course, this is not the case — as they and their luckless viewers soon find out! The chaotic (and usually over-long) movies they produce quickly become a bore, and as a consequence the camera is often laid to rest in a drawer and rarely sees daylight.

What makes all this so sad is that such a situation is totally unnecessary. If only the proud new camera owner realises that he won't get 'professional movies at the press of a button' (as certain ads have hinted), but needs to learn a few basic facts about movie technique, much of this disillusionment can be avoided.

And acquiring this basic knowledge is not really difficult. Movie making is not some mysterious process beyond the understanding of ordinary mortals. True, in the professional sphere it's technically complex, but it doesn't have to be like that; a very great deal can be achieved with the simplest equipment. The point is that the *essential* basic principles are relatively few and can be easily understood by anyone willing to learn.

Just how these principles operate in practice is what we're going to consider first. And I don't think it's an exaggeration to say that once you have grasped them, you're well over half way to making the sort of videos your viewers will be really glad to see.

## Not a branch of photography

The first thing we have to understand is that video movie making is not just another branch of photography. I've actually heard people talking of having been still photographers for a number of years and then 'graduating to movies', as though they were just a more advanced form of the same art. In reality, nothing could be further from the truth, and until we rid ourselves of this notion, we are unlikely to get into the frame of mind that will enable us to make successful videos.

I would go so far as to say that movie making has almost *nothing* to do with photography. Oh yes, both use the same image-forming properties of light and employ a camera as the basic tool, but that's where the superficial resemblance ends because the objectives of a photographer and movie maker are totally different. Not only are they different, but they're also mutually exclusive — which is why it's so important that we get the matter clear in our minds at the start.

With photography, every individual picture is an end product complete in itself. The object — and I'm referring here to those who take their photography seriously — is to say all that needs to be said about the subject in one carefully contrived image snatched from a single moment in time. Many factors may be involved to achieve this — choice of angle, composition, lighting and (often most important) releasing the shutter at just the right moment — but however it's obtained, the resulting picture has got to be a self-contained statement that communicates its message to the viewer as he or she studies it at leisure.

How different this is to a movie which has the added dimensions of movement and time. Now spectators are no longer free to view the scene at their own pace but need to have their attention directed from one detail of the action to another so that they can take them in while they remain on the screen. I used the word *directed* advisedly, and feel that it's worth making the point that a movie director is not just directing the actors and crew during production; there's a very real sense in which he's also directing the audience at the moment they're watching it.

He does this by breaking down the action into a number of different viewpoints, called shots, which can remain on the screen for as long or short a time as is necessary. It's as if he is saying to the viewer: 'Here we have a Western saloon (general shot of saloon). See that group of men sitting in the corner playing cards (closer shot of four men sitting round a card table)? Now watch this — one of them's getting suspicious that there's cheating going on (a very close shot of a player's face as his eyes narrow) and he intends to do something about it (another close shot as his hand moves to his gun). But he's not quite quick enough (a gunshot is heard and the camera switches to a player on the other side of the table holding a smoking six-shooter).'

It's quite likely that these five shots would occupy less than half a minute of screen time, yet they have imparted a great deal of information by establishing a location and developing a dramatic situation. Because of the way the scene has been put together, the viewer knows exactly where he or she is and what's happening. Now if the camera had just stayed with that original general view of the saloon, the chances are that the audience wouldn't even have noticed the poker players in the corner until the shooting started, and so all the drama would have been thrown away.

This, of course, is a very simple and hackneyed example, but it does have the merit of illustrating the most basic principle of movie making: that one shot must lead logically to the next to produce the illusion of an unbroken chain of action. I say *illusion* because playing about with real time is a very important aspect of movie technique; some routine actions have to be

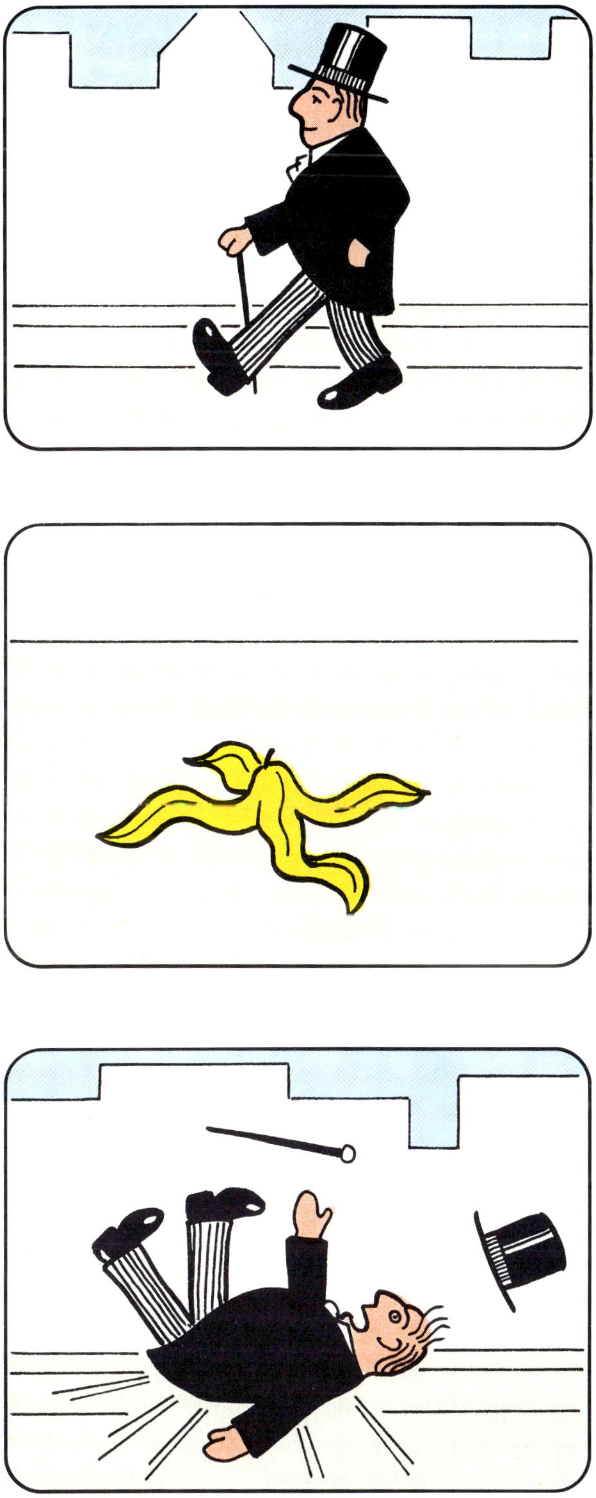

*Cause and effect in the service of comedy. When a shot of a pompous gent it followed by a close-up of a banana skin, the viewer will associate the second image with the first and gleefully anticipate the inevitable tumble. The laugh that this produces will then be all the bigger because the joke has been logically prepared.*

shortened to avoid boredom, while exciting ones are often extended to increase drama and suspense. How to do this without the seams showing is something I shall be covering in detail later on in Chapter 8.

## Different subject, same rules

'Yes, yes,' I can hear some readers saying, 'that's all very well, but I've no intention of making anything like the sort of B Western you quoted, so this doesn't apply to me.'

Ah, but it does! No matter what kind of videos you wish to make — family records, documentaries, dramas or comedies — the same basic rules of movie language apply. Think of it this way: the grammar we use when describing a summer holiday in a letter to a friend is (or should be!) the same as that used by the author who carries off an important literary prize. Of course the style and content are very different, but the fundamental principles on which the words are put together to make sentences are identical.

Even the most matter-of-fact statement needs to be properly formulated, otherwise it will communicate nothing to the reader. If I were to write:

*Day rained Monday all it on*

you would certainly have to guess what I was trying to say. But when the same words are rearranged as:

*It rained all day on Monday*

the meaning at once becomes clear. Order has been brought out of chaos.

So it is with movies. A hotch-potch of shots that have been jumbled together without any regard for logical progression or feeling of continuity will only confuse viewers — a sure-fire way of making them lose interest. Of course, no analogy should be pushed too far, and I'll readily admit that most video shots will convey rather more meaning in isolation than will one written word. But the principle still holds that the fullest meaning — or possibly a completely different one — will only emerge when a shot is given a context by relating it to others in a logical pattern.

## Movie logic

But what *is* the essence of this movie logic that I'm talking about, and just how does it work? It really boils down to three simple concepts: cause and effect, action and reaction, question and answer. This means that one shot carries the implication that it needs to be complemented by the next. Viewers instinctively feel this need (the psychology of good old human curiosity at work here) and expect it to be satisfied. When it is, the separate images blend and achieve a unity unique to the movie medium. Let's just consider each of these concepts in turn.

*Cause and Effect.* To take a time-honoured example from the great days of slapstick comedy, the image of a banana skin on a pavement could be followed by a shot of a portly gent slipping and crashing to the ground. Cause and effect can hardly get more basic than that. But for the purpose of creating comedy, the principle can be elaborated. We could first show the man proceeding on his pompous way, then introduce the shot of the banana skin. The audience will automatically associate this image, and its potential for disaster, with the man in the previous shot; they will anticipate the slippery encounter of foot with skin and start to titter. If we now follow with the shot of the actual fall, the titter will explode into a hearty laugh. This is all in accord with the classic principle of comedy construction often summed up as: 'Show 'em what you're going to do; do it; show 'em you've done it.'

In the realm of family movies, the cause-and-effect concept might be used like this. Shot of little girl smiling happily; shot of adult hand offering a toffee apple; shot of little girl bawling her head off. The message conveyed by these three shots is quite clear: here we have a little girl who hates toffee apples so much that the mere sight of one brings on the tears. But do you remember my saying earlier that the way shots are related can affect their meaning? Well consider this. If we simply reverse the order of those three shots, so that we now show girl crying, toffee apple being offered and girl smiling, they tell a quite different story — here's a child who obviously loves toffee apples.

*Action and Reaction.* The expression as used here hasn't got anything to do with Newton's famous Laws of Motion. It means the reaction of a character to something he or she is presumed to have seen and which the viewer has already been shown in the previous shot. It can range from the horrified expression of the heroine on finding a headless corpse in the boot of her car, to the proud and benign smile of Grandma as she watches her latest grandchild take its first faltering steps.

As we share a common humanity, we tend to react in the same way to certain basic stimuli, so viewers subconsciously find the reaction shot particularly satisfying because it usually confirms their own feelings. This is very useful in a drama because it can strengthen the viewer's identification with a sympathetic character. On the other hand, if the reaction shown is *not* the one they expect or approve of, a feeling of alienation is produced — again, a useful device in a drama for making them hate the 'baddies'.

So powerful is this effect of viewers wishing for a reaction to a shot that has already evoked a response in them, that they will even superimpose their own emotion on to a character wearing a neutral expression. This is one reason why screen acting is a completely different art to performing on the stage. In a movie, a kind of negative acting is often preferable.

*Question and Answer.* Here we are dealing

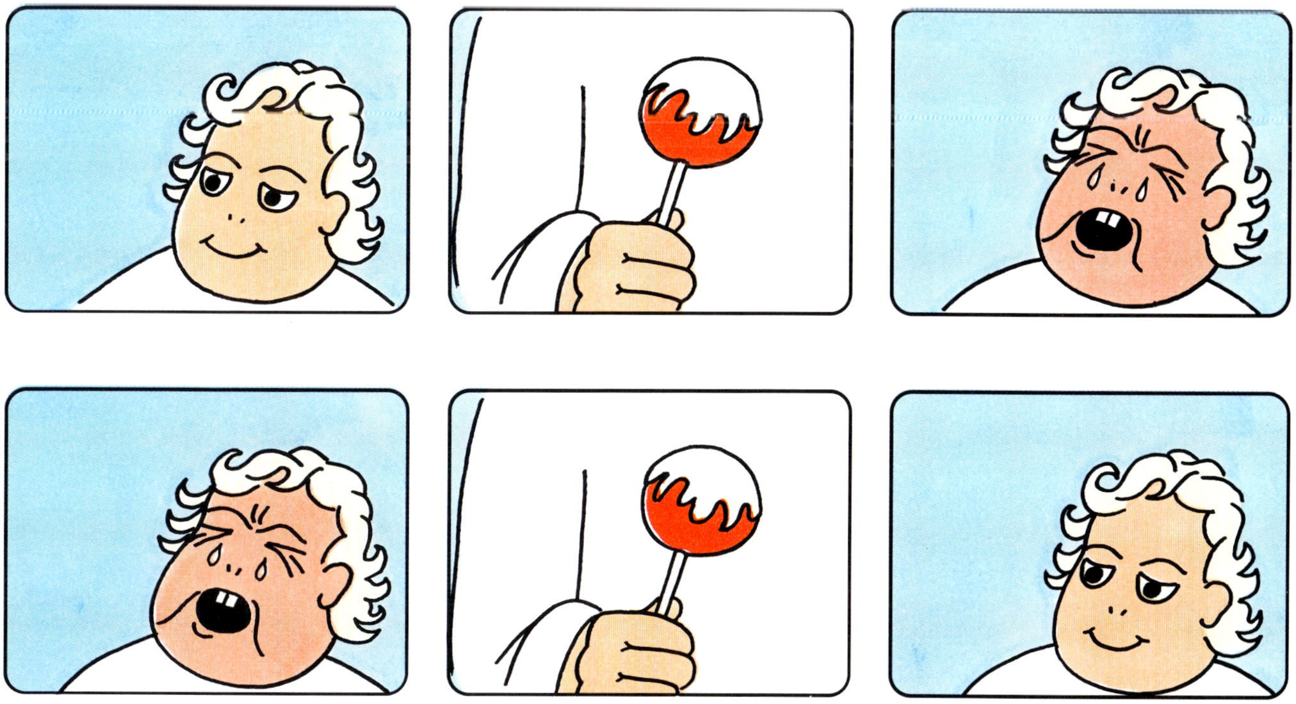

*The same pictures given a new order can tell a different story. Top sequence: child smiling; toffee apple is offered; child crying — this shows us a child who hates toffee apples. Lower sequence: child crying; toffee apple is offered; child smiling — now we see a child who obviously loves toffee apples.*

with the situation where shot A raises a query in the viewer's mind and needs shot B to supply the answer. In many cases it may be a kind of reverse instance of the reaction business I've just been talking about. We see a shot of a man laughing heartily at something happening off screen, and our instinctive reaction is to want to know what he's laughing at. Or again, a wife is shown emptying a bucket of water out of an upstairs window when it's already been established that her husband is weeding a flowerbed immediately below . . .! In both these cases, it's the movie maker's business to satisfy the viewer's curiosity by following with a shot that will answer the question raised.

As you've probably noticed, these three concepts can overlap. There's obviously an element of action and reaction as well as cause and effect in our girl/toffee apple example, and a touch of both these concepts in addition to question and answer in the business with the bucket of water. When this happens, the psychological bond between the shots is naturally that much stronger.

It's of the utmost importance that we remember this need for a feeling of logical progression from shot to shot when shooting our videos. We should always be bearing in mind how the shot we are planning to take will relate to what has gone before and is likely to come next. In other words, we have to acquire the habit of thinking in *sequences*, rather than individual shots. And from this it follows that if you can't see a particular shot fitting into the general scheme of things, then there's little point in taking it.

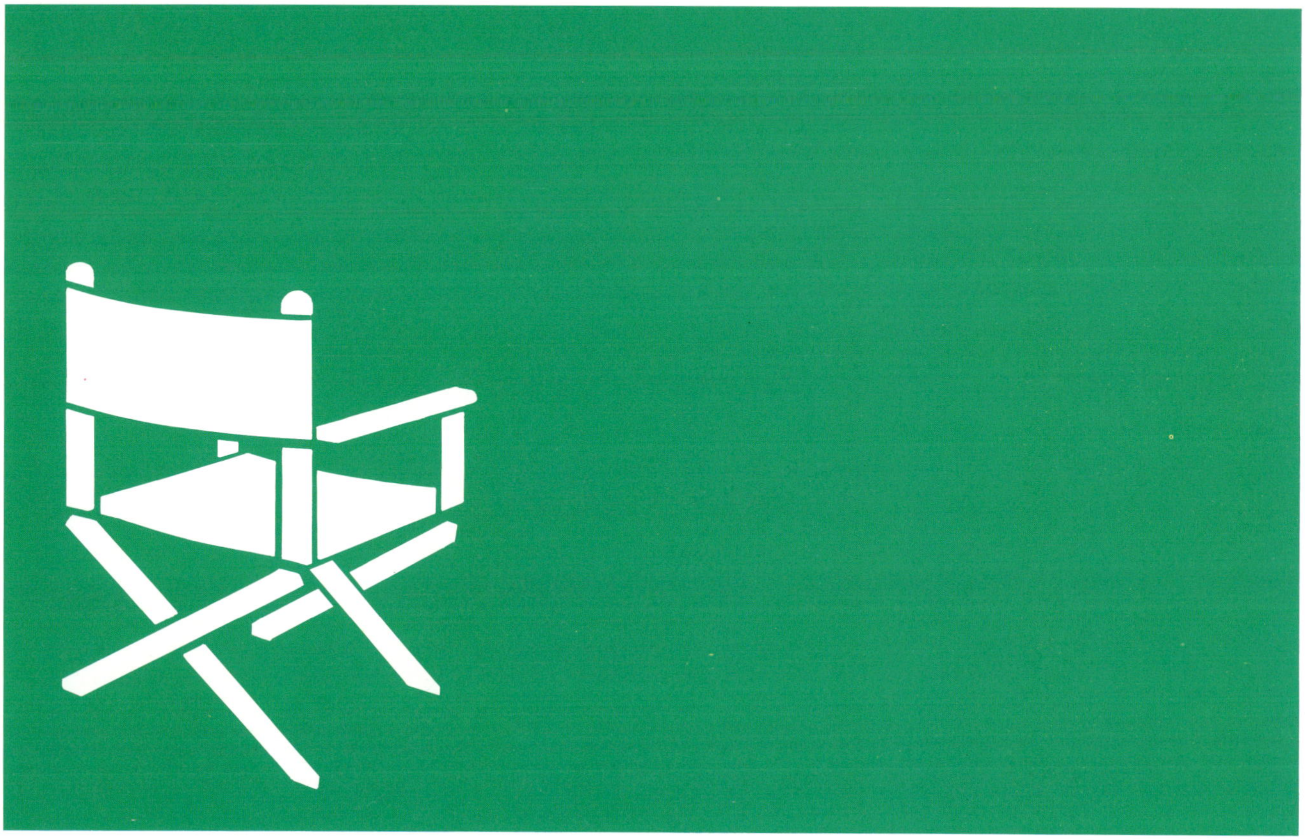

# Chapter 2
# **Exposure**

A modern automatic exposure system mostly does a good job, but there are still times when it can use a little human assistance. This is because such a system is only a mechanism and can't think for itself. It can only measure the *overall* amount of light being reflected from the subject and can't interpret this in relation to the *nature* of the scene. This is of no importance when shooting what are called 'average' subjects — groups in gardens or close shots on the beach, say, where there's not a vast contrast between the lightest and darkest areas — because the system has been set up to handle just this sort of scene.

## Troublesome subjects

Where the trouble starts is with the so-called 'non-average' subject, in which large, but unimportant, areas of the scene are transmitting a great deal more or less light than the main feature of interest. For instance, when a figure is shot from a low angle against a bright sky, the meter will see this background, 'believe' there is a superabundance of light and promptly close down the lens to a very small aperture. This is likely to have the effect of grossly under-exposing the person we are actually trying to record, which means the detail will disappear in murky shadow.

Then there is the opposite case when we shoot a subject against a very dark background — perhaps a white cat in front of a black fence. This time the meter will want to open up the aperture because it's reading a relatively low level of reflected light, so the already light-coloured cat will appear completely washed out through serious over-exposure.

These, then, are the main problems with automatic systems, so what are the solutions? Before going on to those, I ought to say a brief word concerning *f*-stops (also known as *f*-numbers). Still photographers, of course, know all about these, but if you haven't met them before they express the ratio of the focal length of the lens to the effective diameter of the aperture — the factors which determine the brightness of the image being focused on the camera's image sensor.

If this kind of mathematical expression seems off-putting, don't worry; for practical purposes all you need to remember is that any *f*-stop lets in twice as much light as the next higher number. So, for example, *f*/8 gives twice the exposure of *f*/11 and half the exposure of *f*/5.6. Also, when a camera is said to have an *f*/1.6 lens, it means that this is its largest aperture.

Unlike most still cameras, video camcorders usually only have the maximum aperture *f*-number inscribed on the lens, so talk of individual *f*-stop settings is somewhat academic. However, as many camcorders have an electronic viewfinder which shows you what's going on to the tape, manual control can be based on assessment of what actually looks right as you adjust the lens.

But to get back to dealing with those 'difficult' scenes. In the case of the figure against the sky background, all we need to do is open up the lens by the equivalent of about one stop more than that set by the meter. On some cameras, this may be done by pressing what is known as the *backlight button*, which is a feature included for just this situation. Other cameras may have a manual override facility which allows the aperture to be set by hand quite independently of the automatic system.

To put detail back into our washed-out cat in front of the dark fence, we adopt the opposite procedure and simply close down the lens aperture by one stop. In both these instances, the remedial action we have taken will cause the backgrounds to be over- and under-exposed respectively, but this is of no importance because the main subject is now correctly exposed — and that's what really matters.

When making these manual adjustments, watch the electronic viewfinder image closely, and be sure no important detail is lost in shadow or burnt out in highlights.

At this juncture, it might be as well to make the point that the correct exposure for a given subject has nothing to do with how far away it is from the camera. (An exception would be if you were using a camera-mounted video light, but then it's the light-source-to-subject distance that becomes the critical factor.) This means that we can use an alternative method for dealing with 'difficult' scenes to the open-up-one-stop/close-down-one-stop rule of thumb already mentioned.

### Take a close-up reading

Referring back to our cat-and-fence example, we could go right up close and take a meter reading from the animal that would exclude the dark background. Having set the exposure in this way, we lock the aperture at this setting (assuming, of course, that our camera has this facility) and return to our original position to take the shot. However, when using this method, make sure that you're not casting a shadow across the subject while taking the meter reading, as this would obviously cause an incorrect result. If your lens has a long enough telephoto setting (see

Chapter 4) you could zoom in on the cat instead of physically going up close to it.

This method of taking a close-up reading sometimes provides a better result than adjusting the aperture manually while watching the viewfinder because monochrome finders don't always reveal subtle differences in exposure — especially where the matching of flesh tones (which is always desirable) is concerned.

Up to now, we've considered exposure in shots where it's been assumed that the light level remains constant throughout the take. But there are occasions when this is not the case, as when we turn the camera to the left or right (called *panning,* see Chapter 6) to follow a moving subject which may be passing from an area of the

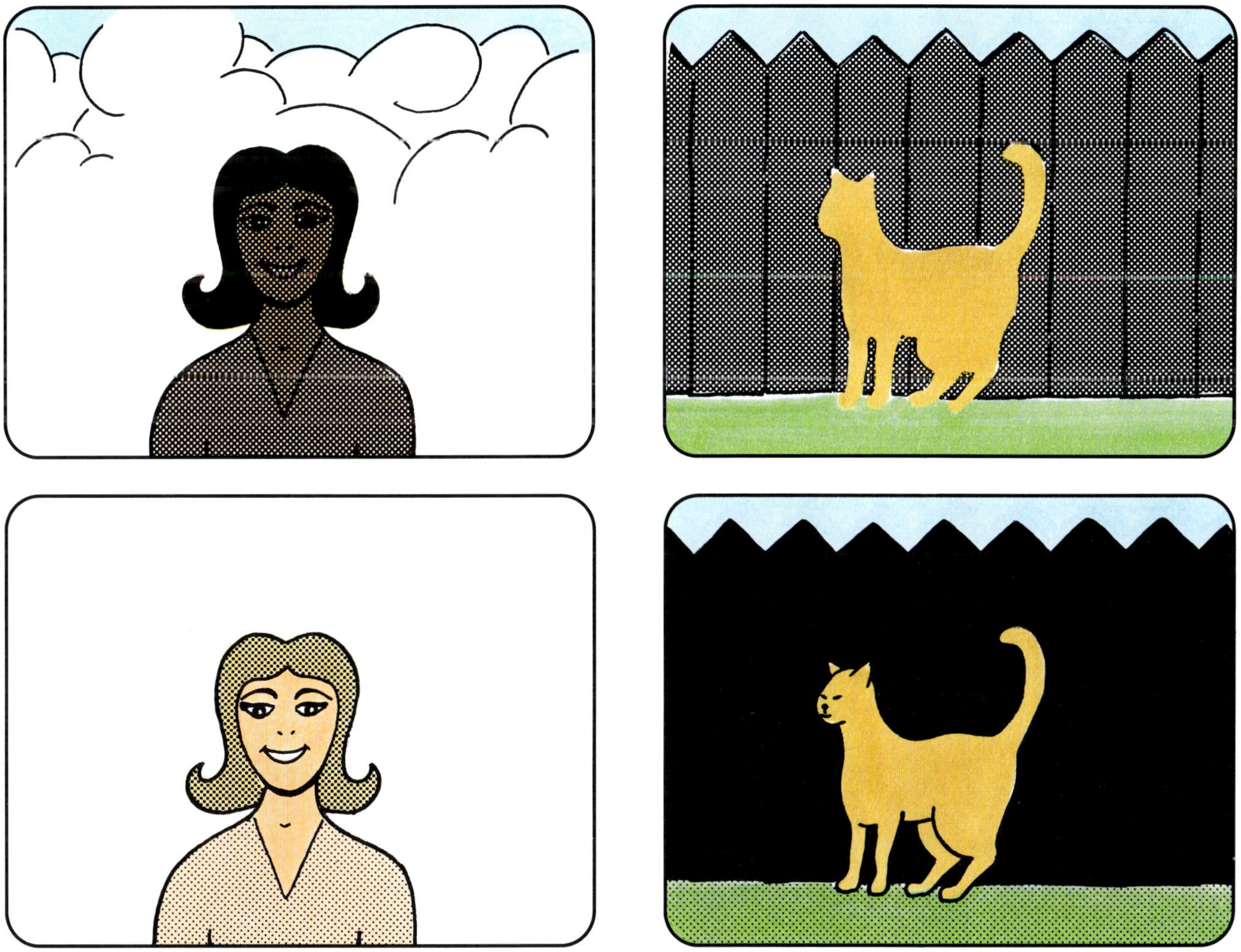

*The top two pictures illustrate what happens when an automatic exposure system is 'fooled' by a non-average subject. In the first example, the large expanse of bright sky in the background has caused the meter to close down the aperture excessively, resulting in under-exposure of the principal subject — the girl. The second example gives the reverse effect: here the dark fence has caused the lens to be opened up too much, so that the white cat is over-exposed. The lower two pictures show the improved results when the lens is opened up and closed down by one* f-*stop respectively. In each case the main subject is now correctly exposed, although the detail in the backgrounds has been lost.*

scene with a dark background to one with a lighter background. So doing will cause a fully automatic system to react to the increase in light coming through the lens and close down the aperture during the course of the shot. This will, in turn, affect the exposure given to the main subject we are following, and result in some darkening of its colours (the colours would, of course, lighten in the opposite situation where the camera was panning from a light to a dark background). Such colour changes look rather unsatisfactory when the shot is played back, especially if the subject is human and flesh tones are involved.

If our camera has no aperture lock, there isn't very much we can do about this except, perhaps, use two separate shots instead of the pan, allowing the subject to move out of one and into the next so that the exposure is correctly adjusted for each. Obviously the colour values will still have made the same changes between the shots, but this won't be as noticeable as when it takes place on screen.

*When exposure is being influenced unduly by an over-bright or dark background, one answer is to take a meter reading from the subject itself — provided your camera has an aperture lock. The camera operator advances to position B and directs the camera at the subject from close range, making sure that his own shadow is not falling across the subject. The aperture is then locked on this exposure setting, the camera operator returns to position A and takes the shot.*

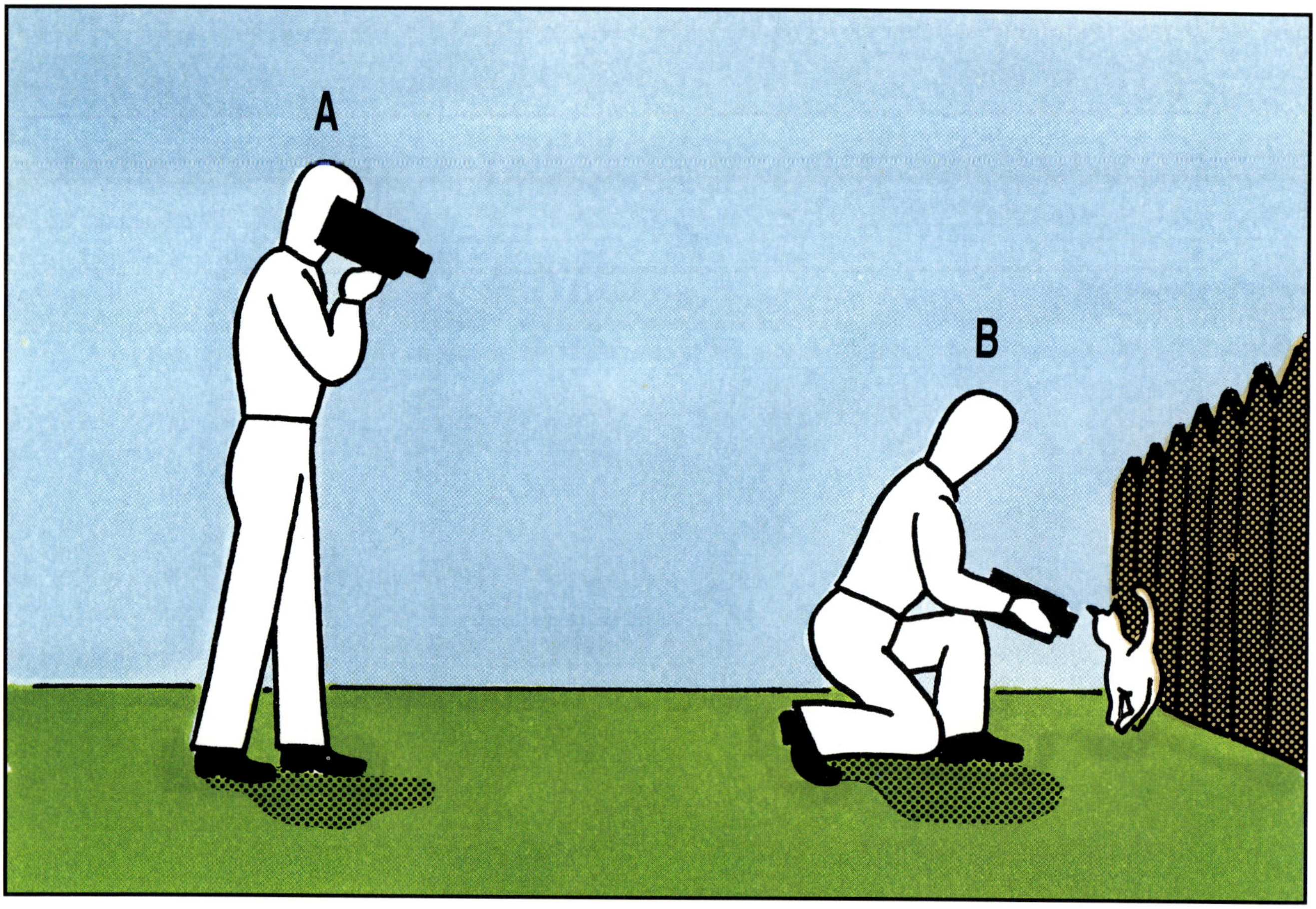

*Shooting scenes in snow can often be tricky because so much light is reflected that the automatic exposure system closes the lens to a very small aperture. This usually means under-exposure of figures seen in medium or close shots and manual adjustment needs to be made to compensate. However, in a general long shot, where the snowscape itself is the subject, the auto meter can take care of the exposure quite happily. Note here how the texture of the snow is brought out by the cross lighting.*

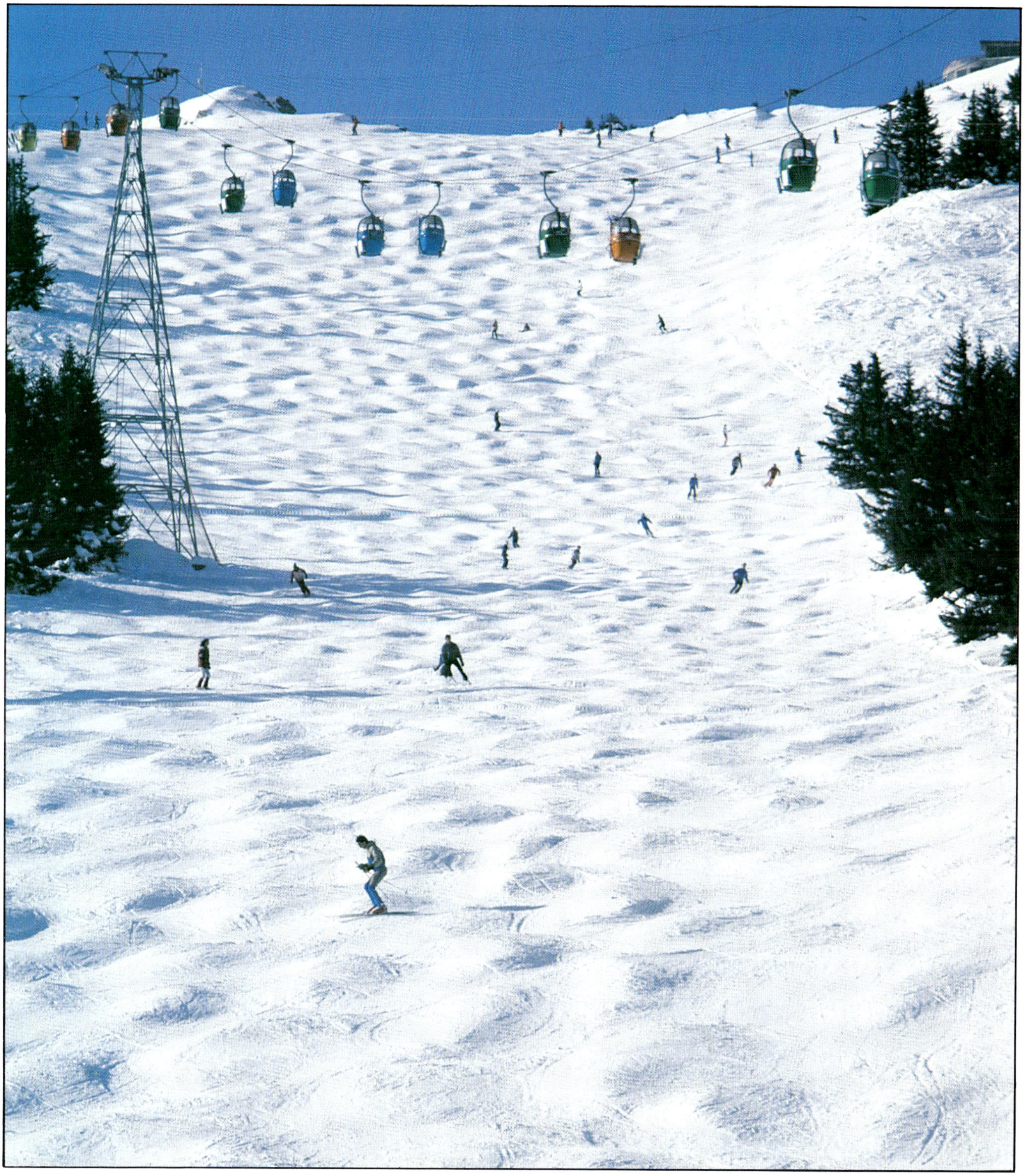

However, if we have manual control of the aperture, the thing to do is set the exposure at the start of the shot and lock it on this setting, which will ensure that the exposure of the main subject will remain constant as it moves across the scene. Perhaps I should point out here that if the subject *itself* is moving out of (or into) shade, then the metering system should be left free to adjust for the varying exposure needs accordingly.

Mostly when our camera is assessing exposure, it's dealing with light that's being reflected by the subject, but there are occasions when the light coming into the lens is *transmitted*. This happens when the subject itself is, in effect, the light source — fires, illuminations, stained glass windows and so on. But this doesn't really make any difference as far as the meter system is concerned; it will still measure the amount of light available and try to set the aperture as though for the notional average scene.

This can sometimes result in over-exposure because of the 'dark background' effect mentioned previously. For instance, if a neon sign occupying a relatively small area of an otherwise black sky were being shot, the meter would throw the lens wide open in its struggle to get 'average' exposure. On the other hand, a bright display which completely filled the frame might cause the aperture to close down a little too much. Before doing any important shooting of such subjects, it would be as well to experiment to see just how your camera behaves in these unusual situations.

### Low-light shooting

While on the subject of exposure, it might be appropriate to say a few words about the *low-light* shooting facility offered by many camcorders. This works by electronically boosting the video signal (it also helps if the lens has a large maximum aperture — usually *f*/1.2), and allows you to shoot in relatively low levels of available light — typically down to 10 lux (lux is a metric unit of light intensity).

This means you can record under lighting conditions that would normally be impossible with an ordinary camera, such as indoors without the aid of any special photographic lamps. But like most convenient things, such an advantage can only be achieved at a price — in this case, a lower standard of picture quality. The laws that govern electronic image reproduction decree that such signal-boosting will increase visual 'noise', so that the picture has a coarse, grainy look about it. There is also a likelihood of increased 'comet-tailing' — the residual trails that follow highlights in dark scenes.

Another point about low-light shooting is that when most lenses are opened up to their maximum aperture, their performance tends to be a little 'soft', which means the image they produce is not quite as sharp. So pictures taken under low-light conditions with wide-aperture lenses will not have the clarity and 'bite' we expect when shooting under more normal circumstances.

To draw attention to these limitations is not to knock the low-light shooting facility; it can be very useful in many situations where it would be impossible to obtain coverage in any other way. But all the same, it must be said that when we can have everything under our own control and plenty of time to take each shot, it's always best to light indoor or atmospheric dark scenes properly (see Chapter 10) so that we obtain the best possible image quality in the finished production.

# Chapter 3
# **Focus**

Even those with no practical experience of photography have heard about focus and know it has to do with getting a sharp image of the subject in the final picture — either in a print or on a screen. Newcomers to video may think it's a complicated business that they'll find difficult to handle, but happily such fears are quite unnecessary.

It's a fortunate fact of life that, under most circumstances, focus is much less of a problem in movie making than it is in still photography. But before we consider the practicalities of the subject, a little painless theory about how lenses form an image might be useful.

Any subject is recorded photographically (and I'm using the term in its widest sense to include electronic methods) by the light reflected from it. This reflected light can be regarded as coming from an infinite number of points on the subject, each of which is bouncing off the light rays — duly modified as to colour and intensity — in an infinite number of directions. Assuming we have our camera pointing at the subject, some of these rays will pass through its lens. Now let's consider what happens to the rays from just one light point which we'll call X.

All the rays from X which strike the lens are bent or *refracted* by it (see accompanying diagram where five rays are illustrated for the sake of clarity) in such a way that they converge in a cone shape on the far side of the lens. If the surface of the image sensor of a camcorder or TV camera, which we'll call AB, is placed at the tip of this cone, Y, a sharp image of point X will be formed upon it. But if the sensor surface is moved towards or away from the lens (in practice, of course, it is the lens that would be moved) the image of X would no longer be a single point but a circle. When this circle — known technically as the *circle of confusion* — exceeds a certain size, the image is said to be out of focus, which simply means that it appears unsharp and fuzzy.

I should, perhaps, make it clear that for the sake of this illustration I've represented the lens as a single piece of glass, whereas in practice a modern zoom is made up of several elements. However, the basic optical principle is the same.

The distance at which the lens must be set from the sensor to produce a sharp image varies according to how far the subject is from the lens, so theoretically there can be only one 'correct' focus setting for any given subject-to-camera distance. But don't worry — this is not as restricting as it might at first seem because we can make use of an invaluable phenomenon called *depth of field.*

### Depth of field

When the camera is focused on a subject at a certain distance, the depth of field is simply the range from a point in front of the subject to a point beyond it within which everything looks acceptably sharp. 'Acceptably sharp' is a non-scientific way of saying that the circle of confusion does not exceed a value that is calculated in relation to the size of the image being formed.

*How a lens forms an image. Rays of light from X, a point on the subject, which pass through the lens are refracted and come to a focus on AB, the camera's image sensor, at Y. If AB is moved towards or away from the lens to the positions indicated by the broken lines, the image of X will no longer be a point but will spread into a circle and so be 'out of focus'.*

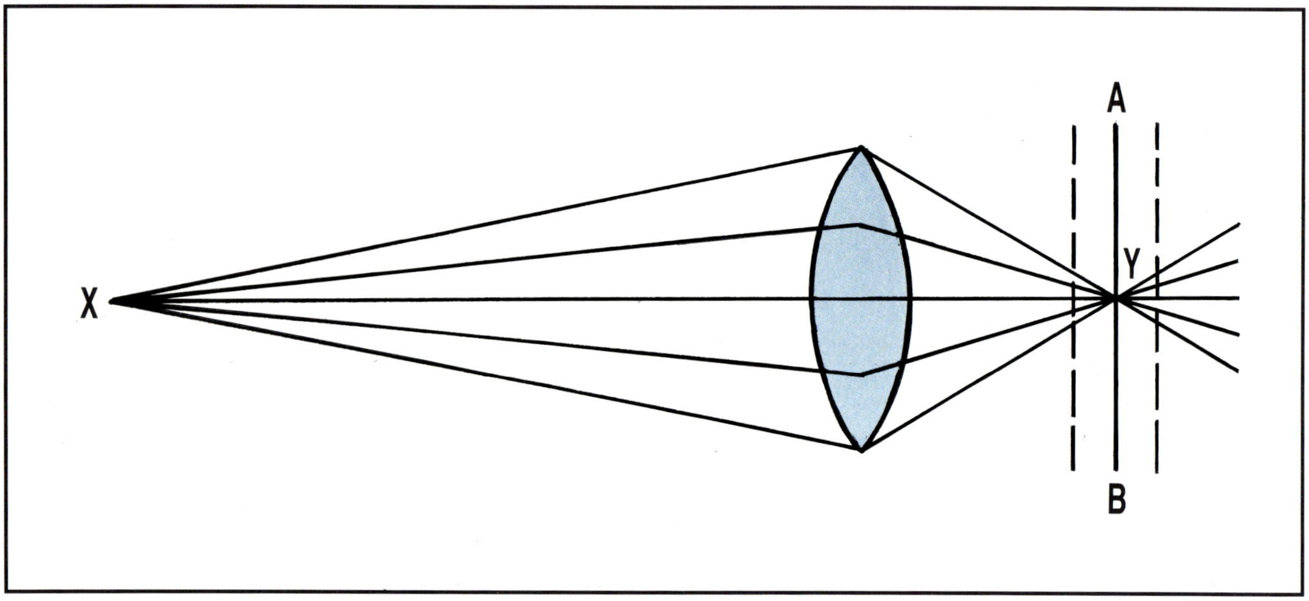

*One of the factors which affect depth of field is focal length (see Chapter 4). Here the shaded areas indicate the comparative depth for wide-angle, normal and telephoto settings of a zoom lens when focused at the same distance and working at the same aperture.*

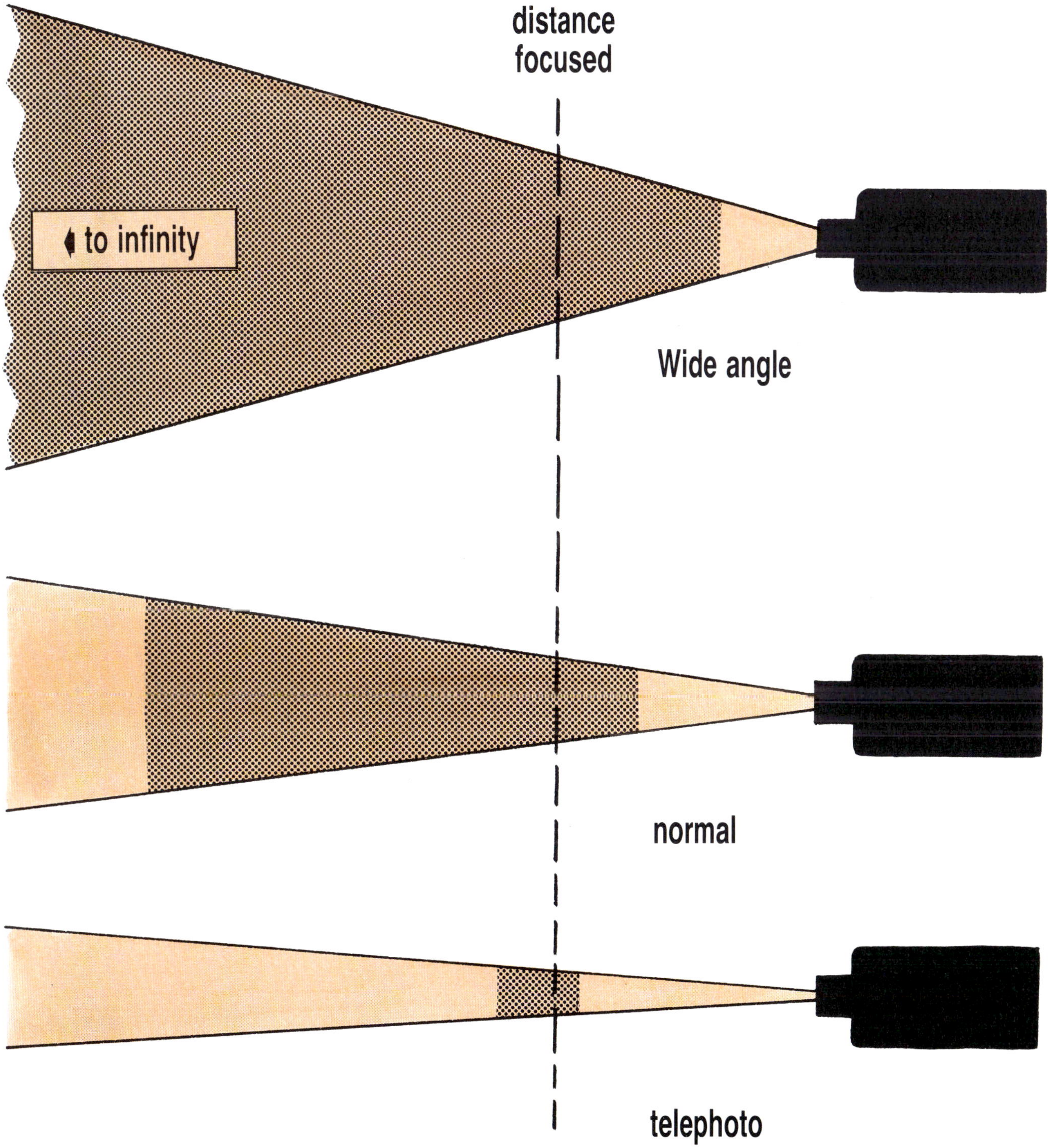

# 3

Three factors affect depth of field, and it's very important that we understand them and bear them in mind when shooting. Depth of field *decreases* as:

(1) The lens aperture is increased (to give more exposure)
(2) The focal length gets longer (as when zooming towards telephoto)
(3) The lens is focused at shorter distances (for subjects near to the camera).

This means that on a bright summer's day, with the zoom set towards the wide-angle end of its range (say, 12mm) and focused on 6ft, we would have a depth of field from less than 3ft to infinity (a theoretical point in the far distance from which the proceeding light rays are regarded as being parallel). On the other hand, if the weather is dull and you're shooting at maximum telephoto on a relatively close subject, then the depth of field will be extremely limited and accurate focusing is essential.

The 'normal' setting of a camcorder zoom would be somewhere around 15mm. When we wish to zoom in on a subject, we have to increase the focal length (see Chapter 4) which causes the depth of field to decrease dramatically. This means that if the lens has not been carefully focused beforehand, the picture will dissolve into a horrible blur as we zoom into our telephoto close-up. It needs to be emphasised that this is one of the major traps for the inexperienced. However, zooming back to a wide-angle poses no such problem because, as we have seen, depth of field increases when the focal length is reduced, as it has to be for wide-angle shots.

### Fixed-focus working

As I've explained, when shooting in good light, at a 'normal' zoom setting, with the lens focused at about 6ft, we have all the depth of field we need, so under these conditions we can use our camera in a *fixed-focus* mode — that is to say, we don't adjust the lens for changes in camera-to-subject distance. This is a good way to shoot fast, catch-as-catch-can action when speed is of the essence.

But when working under conditions of limited depth of field (in poor light and/or at the telephoto end of the zoom) we must, of course, focus the camera lens carefully. This is done by turning the focusing ring on the lens and observing the sharpness of the image in the viewfinder (assuming it's an electronic one and shows what is actually being recorded).

The main point here is that visual focusing should always be done with the zoom set to full telephoto for maximum sensitivity. Once the lens has been focused, you zoom back to the focal length setting needed to frame the image properly before taking the shot. When focusing, concentrate on some part of the scene with clearly defined points or lines and get these as sharp as possible — if there's any lettering visible, focus on this.

On some video cameras, the video level wave form can also be used as a focusing aid; this works on the principle that the sharper the picture, the greater the contrast and so the more 'spiky' the wave form.

### Auto-focusing

You may be wondering at this stage why I've said nothing so far about the auto-focusing cameras available. Surely they'll solve all focus problems for us? Well, not quite. Marvels of ingenuity though they are, they can't think and interpret the subject. Most of them work by setting the focus from the object that seems to dominate the centre of the composition, which in most cases would probably be what you want. But then again, it may not be.

Perhaps you wish to give a feeling of depth to the scene by having a tree in the foreground with the main subject beyond it, or maybe you want to shoot someone walking along behind railings. In either of these cases, the chances are that the camera would focus on the foreground feature instead of on the wanted subject further away.

*Impressive screen-filling close-ups of small subjects can be obtained by using the macro facility included on many cameras or a supplementary close-up lens. When working in close, there can sometimes be trouble with camera shadow, but here the operator has neatly sidestepped the problem by backlighting the subject, which also gives an added richness to the colours of the flower.*

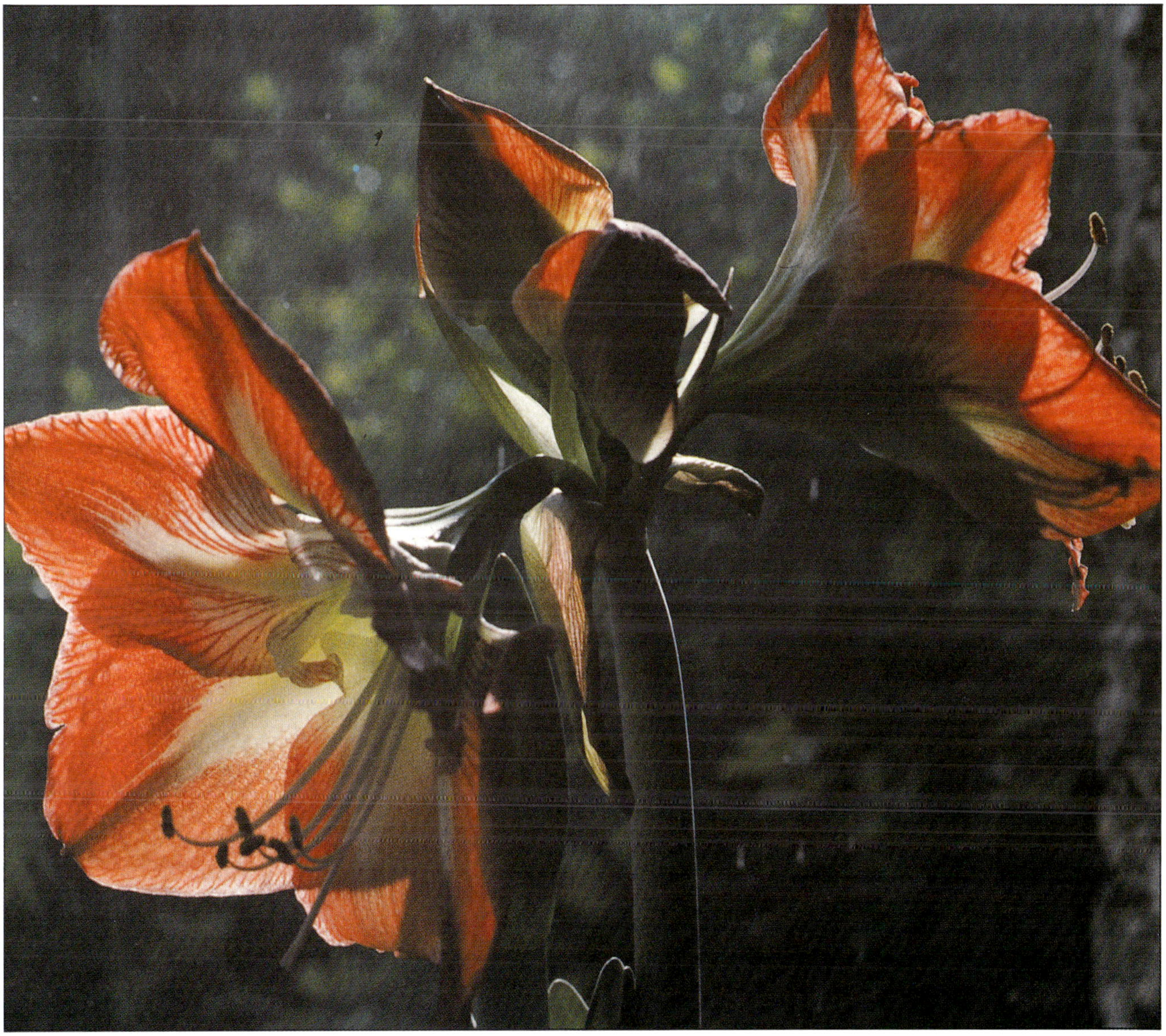

Also, some auto-focusing systems tend to get 'confused' by subjects with little contrast or without vertical lines, subjects behind glass, reflective surfaces and when there are bright lights behind the subject.

So for these reasons I think it has to be said that the facility of auto-focus is not an unmixed blessing, and that one really needs a camera which also has manual override as a second line of defence. Certainly it's not a comparable boon to the automatic exposure feature discussed in Chapter 2.

When talking about depth of field earlier, I may have given the impression by implication that maximum depth is what we always want. This is far from being the case because there are occasions when *differential* focusing — that is to say, we only have a subject at a certain distance sharply defined — can be a very useful device. Perhaps we want to concentrate the viewer's attention on an individual in a crowd, or shift interest from one character to another by *pulling focus* (adjusting the lens to a different distance setting) during the course of a shot.

As we have already seen, the shorter focal lengths used for 'normal' shooting in good light give considerable depth of field, so if we want a differential focus effect we must use a telephoto setting and place the camera further away from the subject (unless a more restricted view is the intention). If we also wish to enhance the effect by using a larger aperture while maintaining the correct exposure, neutral density filters, which reduce the light entering the lens without affecting colour, can be used for this purpose.

### Getting in close

Many cameras with a zoom lens also have what is called a *macro* setting. This can be used to shoot small objects (stamps or transparencies, for example) at extremely close distances — occasionally even touching the front element of the lens. So in some cases this means you can fill the screen with an object little more than an inch wide!

When a lens is switched into its macro mode you lose the use of the zoom as such, but this is a relatively small price to pay for the exciting giant close-ups you can get of, say, insect life. The main point to watch when shooting at such close range is that the camera doesn't cast a shadow on the subject. Obviously this is no problem when recording transparent subjects which can be lit from the back. Also, when working at such close distances there is very little depth of field so you need to be particularly careful about focus.

An alternative method of close shooting, if your camera doesn't have a macro facility, is to use a *supplementary* close-up lens. For example, a supplementary lens of 19½in (+2 dioptres) enables you to work at this distance from the subject (measured from the front of the supplementary lens) with the camera's own lens *set at infinity*. When such a lens is fitted, the camera's zoom is left available for zoom effects.

The camera's electronic viewfinder can, of course, still be used for visual focusing when a supplementary lens is being used, and by turning the camera lens focusing ring in the appropriate direction, you'll find that you can work at even closer distances than the focal length of the supplementary lens.

# Chapter 4
# Facts about Focal Length

The focal length of a lens is the distance at which it needs to be set from the image sensor of a camera to focus an object at infinity. This distance is usually expressed in millimetres, and the zoom lens of a video camera has its variable focal length quoted as a range from the minimum to the maximum setting, say, 9-54mm.

You're possibly thinking that this all seems very straightforward and may be wondering why we need to concern ourselves with the question of focal length anyway; after all, provided the lens is forming a decent image does it matter very much what focal length is being used to do it? Well, yes it does actually, because the focal length chosen affects the 'look' of the recorded subject in a number of fundamental ways, so it's important that we should understand what result a given setting will produce.

In the days before zoom lenses were standard issue, amateur movie cameras could be fitted with three basic types of lens: *normal, wide-angle* and *telephoto*. Many of these pre-zoom cameras

# 4

had two or sometimes three lenses mounted in a turret which allowed a different one to be swung into position at the twist of the wrist.

But obviously this arrangement couldn't compete with the convenience and versatility offered by the zoom where the user had an infinite number of focal lengths available at his finger tips. A modern zoom can of course be set at the equivalent focal lengths of the former normal, wide-angle and telephoto prime lenses, so we'll consider the effects produced under these three basic headings.

### Normal setting

This setting, as the name implies, is used for most normal shooting where the intention is to represent the subject in as natural a way as possible, and where there are no practical difficulties about placing the camera in the right position to take the shot. By representing the subject naturally, I mean without exaggerated perspective effects (that is, portions of the subject nearest and furthest from the camera do not look unduly large and small) and the overall impression approximates to what the eye would expect to see in real life. To achieve this natural look, the zoom might be set at around 15mm.

There will be occasions, however, when although we might wish to have the natural look, practical shooting conditions (we can't place the camera at the ideal distance) make this impossible to achieve because they force us to use a focal length setting that's shorter or longer than the normal one.

### Wide-angle setting

Perhaps we are trying to take a general shot of a

*The focal length of a lens is the distance at which it must be set from the image plane to focus light rays coming from infinity. This diagram illustrates a single-element lens but the principle is just the same for the more complex lenses used in video cameras.*

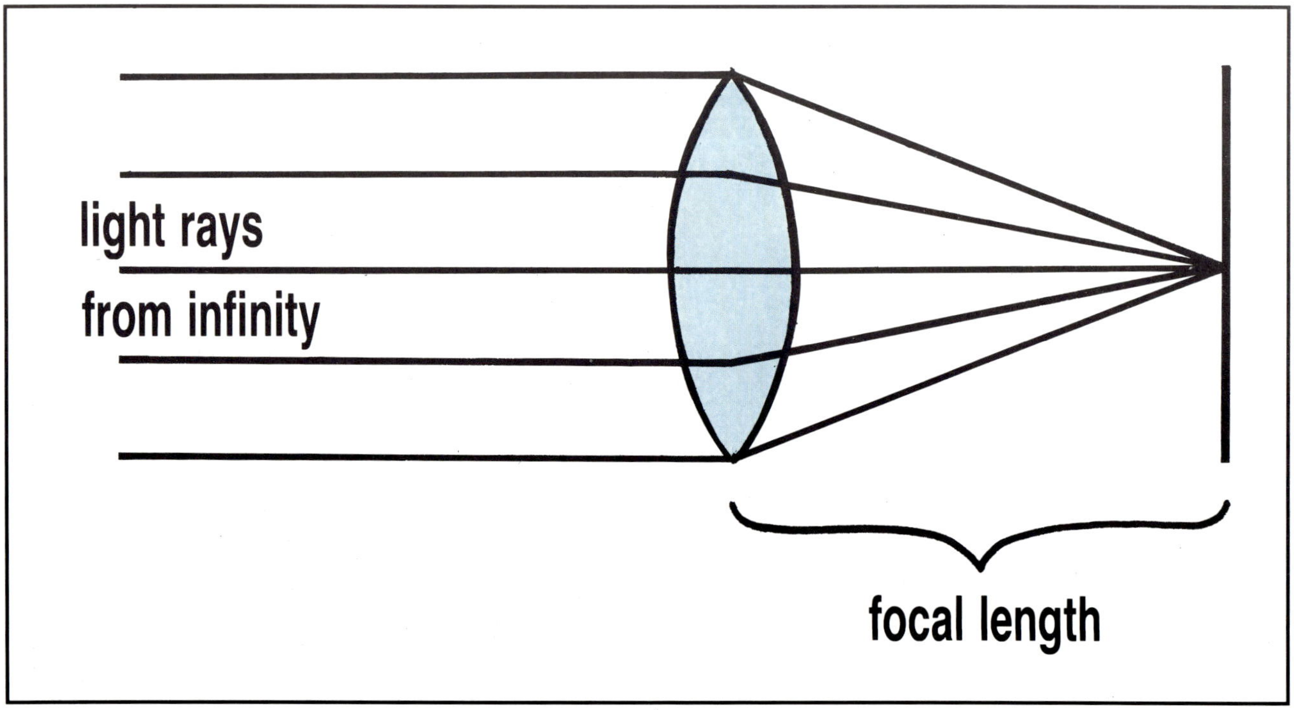

picturesque old town hall standing on one side of a square, but the buildings on the opposite side prevent us from getting far enough back. This is where the wide-angle setting of the zoom would help. When the focal length is reduced to 9mm, the angle of view 'seen' by the lens is widened, so allowing more of the subject to be included in the same frame area.

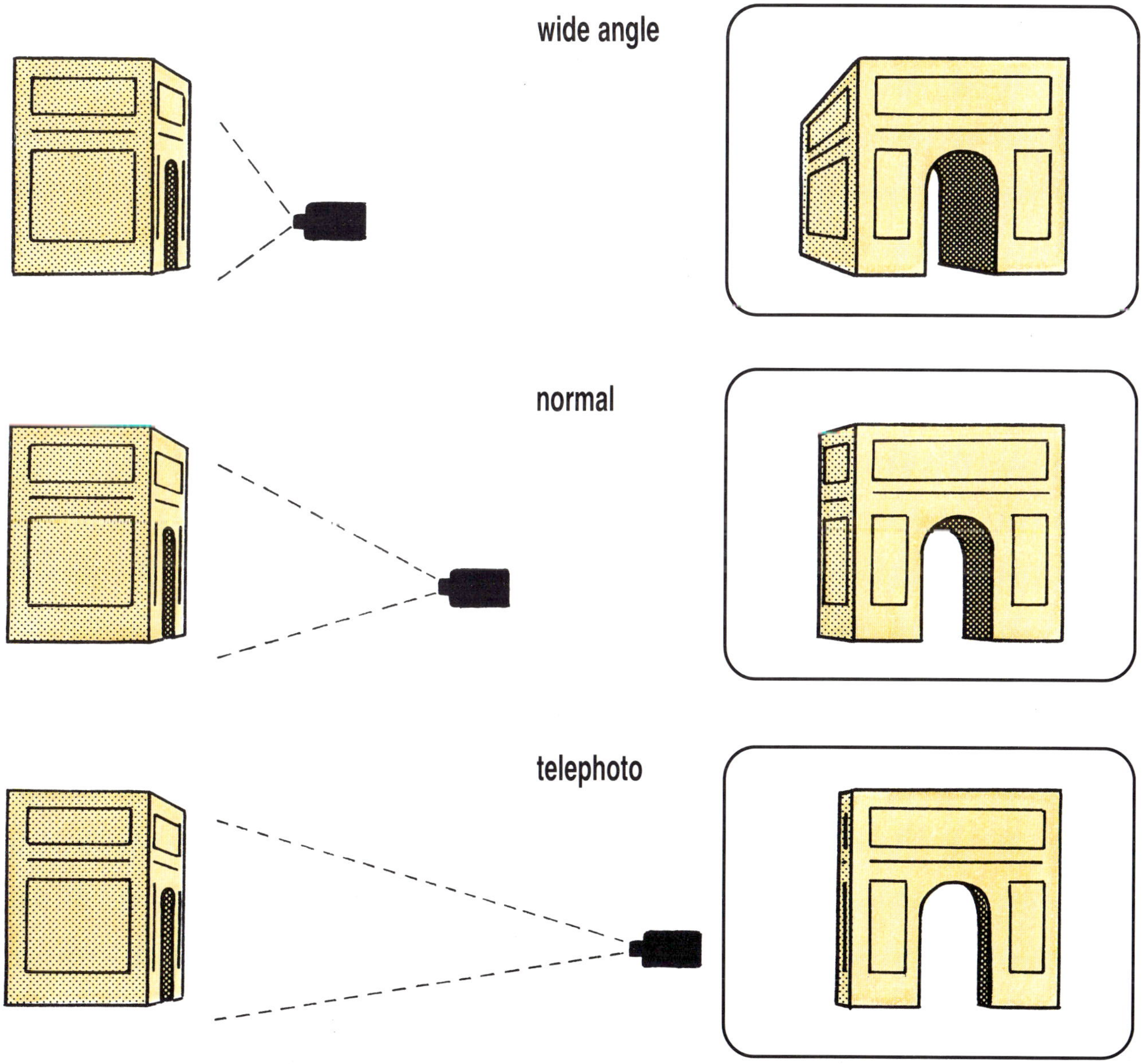


*Focal length and its effect on the 'look' of the subject. When the camera position is adjusted to record the same size image of the arch in the frame, a wide-angle setting gives the impression of exaggerated perspective, a normal setting the more 'natural' view usually seen by the human eye, and a telephoto setting makes the perspective look 'bunched up'. (The camera size and camera-to-subject distances are not, of course, to scale in this illustration.)*

# 4

All very simple. But using a wide-angle setting does have another effect on the look of the subject as recorded besides just allowing more of it to be squeezed into the picture. The perspective will appear to be 'stretched' so that any depth apparent in the subject will seem greater than it would if a normal focal length setting had been used, and objects near to the camera will look disproportionately large in relation to those that are further away.

To be absolutely correct, at this point, I should make it clear that the perspective in a scene is not actually affected by the focal length of the lens in use, perspective being controlled by the *distance between camera and subject.* As you move closer to the subject, perspective is exaggerated and as you go further away it is minimised.

However, if a subject is to occupy the same proportion of the frame area in a wide-angle, a normal and a telephoto shot, you obviously have to get closer to it with the wide-angle and further away with the telephoto. So in practice it's perfectly natural that we associate exaggerated perspective effects with a wide-angle lens setting and a relatively 'flat' (foreshortened) look with a telephoto setting.

Hence, if you take a close-up of your girl friend with a wide-angle setting, you'll have to get very close to her and the resulting 'stretched' perspective will be far from flattering, increasing the size of her nose and narrowing the set of her eyes. So keep well back and make her face fill the frame by adjusting the zoom to telephoto. On the other hand, if you're making a horror movie the wide-angle's distortion of features could be just what you need.

If a wide-angle setting can give the impression of distorting space, then a moment's thought will show that it must also have some effect on movement taking place within that space. Because the distance between near and further points has apparently been increased, it follows that an object travelling from one to the other will seem to be moving faster. So action towards or away from the camera in a wide-angle shot has a speeded-up look. This can be turned to practical advantage and used as a means of creating additional excitement in action sequences.

### Telephoto setting

After taking our general view of the old town hall, we might want a close-up of the clock at the top of its tower. Obviously we can't get physically near enough to it with the camera to take the shot, so we set the zoom lens at the telephoto end of its range (say, 54mm) which should enable us to fill the frame with the clockface as we wish.

With a fairly flat subject, such as the clock, the result would look pretty much the same as if the camera had actually been moved in close to take the shot, but this would not be so if the scene had perceptible depth as well. As might be expected, the effects are the opposite to those we've already noted in wide-angle shots.

It's a general characteristic of telephoto shots that the perspective has a 'bunched up' look about it or, to be more scientific, it's *foreshortened.* We've all seen those head-on shots of horse races where the runners seem to be on top of each other and apparently making little forward progress as they thunder down the track.

A general point to bear in mind when taking telephoto shots is that a tripod or other device for steadying the camera should always be used. This is because as well as increasing the image size, the long focal length also has the effect of magnifying lateral and horizontal movement of the camera relative to the subject, so any shake or unsteadiness can be very distracting in shots of this type.

Another interesting telephoto effect is the 'growing' background feature which can sometimes be used for a symbolic purpose. Suppose, for example, we have a character who, at an early stage of the video is seen looking over a gate at a windmill in the distance. Later on in the movie, he returns to the same spot and once more gazes towards the mill. However, between the two visits, the mill has become an important factor in his life and we want to suggest this visually by making the mill loom larger in the picture. Obviously, there's no way we can move the mill nearer to the gate physically — but we can move it *optically*, and this will do just as well.

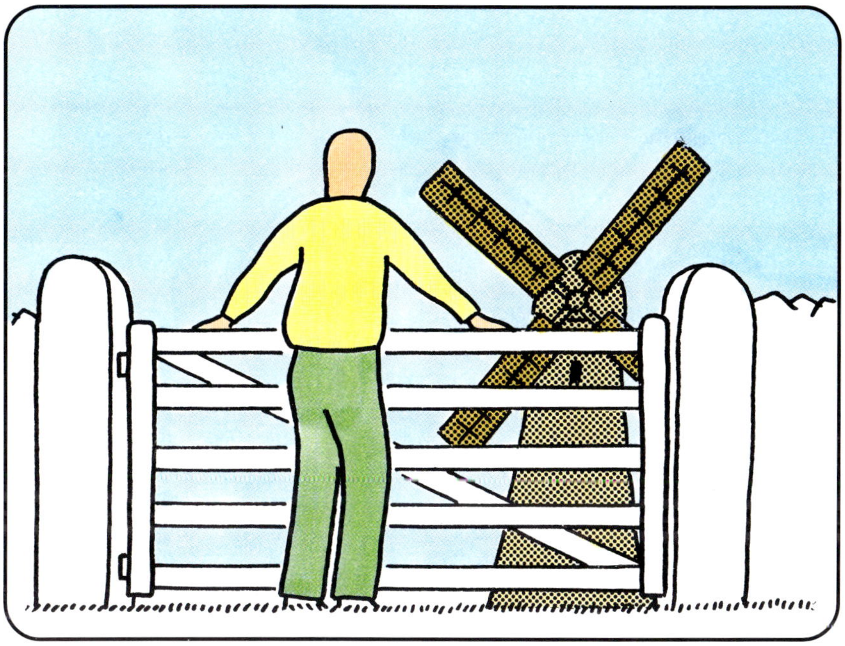

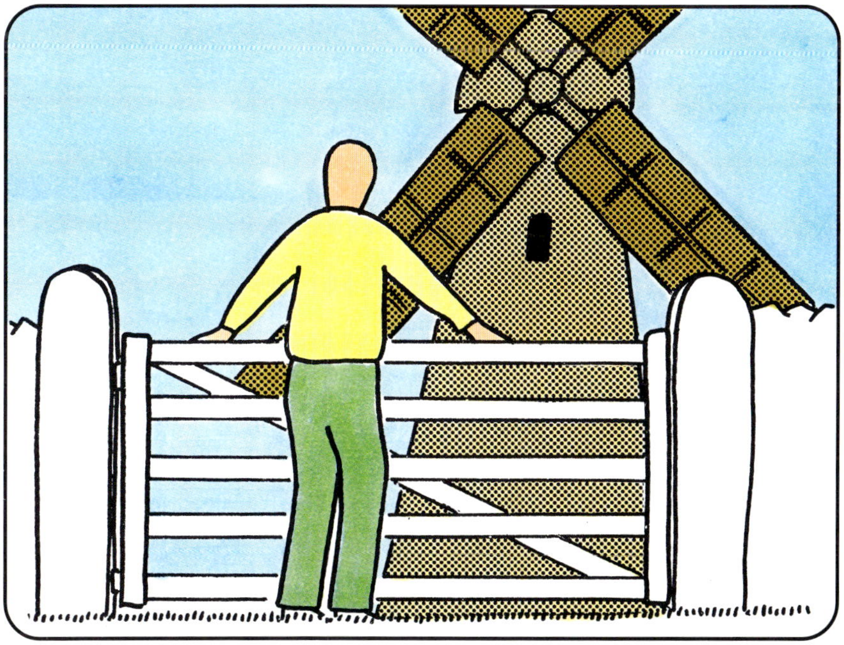

*Making a windmill 'grow'. The top sketch illustrates a shot of the scene taken with the lens on 'normal' setting. The lower sketch shows the same scene taken on a telephoto setting with the camera moved further back to retain the same image size of the man at the gate. The effect is to make the windmill loom much larger in the frame, while the only noticeable change in the foreground subject is the flattening of the perspective of the gateposts.*

The two shots would be tackled like this. For the first we would position our character by the gate and compose the scene so that he and the mill were in their correct relative positions. This composition would then be taken with the lens at a *normal* (or possibly slightly wide-angle) setting. For the second shot, we switch to telephoto and move the camera until the figure and the gate are the same size in the frame as in the first shot. However, the telephoto setting will have brought the mill relatively much nearer so that it now occupies a larger proportion of the frame — which is what we were after.

## Zooming

As I said earlier, a modern zoom can be regarded as a lens with an infinite number of focal lengths built in, any one of which can be instantly selected by means of a lever, ring or motorised control. This adjustment from one focal length to another can, of course, be made while the camera is actually recording which produces the familiar effect known as *zooming*. It might be worthwhile pausing for a moment to consider more closely just what result this zooming action produces in the picture.

A zoom is *not* the same as a tracking shot (see Chapter 6) where the camera is actually moving towards or away from the subject. The spectator's viewpoint may have the appearance of moving, but details of the scene at whatever distance from the camera increase or decrease in size at the same rate during a zoom in or out. This is because there are no motional perspective changes of the kind you would get if the camera were actually travelling.

Zooming, because it's easy to do and has a somewhat hypnotic effect on the cameraman as he looks through the viewfinder, tends to be overdone by newcomers to movie making. It's undoubtedly a useful technique but one that needs to be used with discretion and only with some definite purpose in mind. For example, you would zoom in to draw the viewer's attention to some important detail in the scene, or zoom out from a close-up on an object to reveal the context in which it's placed. But what you should avoid at all cost is the 'tromboning' effect of zooming in and out for its own sake. It's a good guiding principle that if you have any doubts about whether or not to zoom during the course of a shot — then don't!

5

# Chapter 5
# **Image Size (Types of Shot)**

Nowadays, we're all so familiar with the way images of various sizes are interwoven during the course of a movie sequence, that it's hard to believe there was ever a time when a mixture of long shots, medium shots and close-ups wasn't normal practice.

But there certainly was. For the first half-a-dozen or so years of the cinema's existence (it is, of course, from the cinema that TV and video have taken their visual 'language'), the pioneers simply set up their cameras to get in as much of the scene as was practicable and cranked away for as long as the action lasted — or until the film ran out. Even in the early primitive dramas, stage practice was followed: instead of the rising curtain there would be a fade-in on the scene, which would be a general view (long shot) of the sort you would have from a stall seat half-way back in a theatre. All the necessary action in that location would then be performed, and there would be a fade-out before moving on to the next scene.

# 5

However, what worked in the theatre, where the drama can be sustained by dialogue and the physical presence of the actors, was soon found to be hopelessly inadequate for the silent cinema. In a play it doesn't matter that the audience get a fixed general viewpoint of the action because any details they can't easily see for themselves, but need to known about, can be covered by a line of dialogue.

'Look — I've found a ring,' a character might say as she picked up some object too small to be seen beyond the front row of the stalls. But what could you do in a silent movie? Have an explanatory title — 'Jane finds a ring.' — perhaps? Functional, but not very dramatic. However, the penny soon dropped: *why not move the camera in closer so that the audience would actually be able to see the ring being picked up?*

Why the history lesson, you may be wondering? Well, it's just that I want to stress the point that when it comes to the basics of movie making, nothing really changes — technical developments such as sound, colour and video notwithstanding. The fundamental needs of visual storytelling, which caused the pioneers to forge these classic techniques in the first place, are just as pressing today whether we want to make an epic about the end of the world or simply record young Tommy's efforts to build his first sand-castle. The rock-bottom guiding principle is to show viewers *what* they need to see *when* they need to see it, and with the appropriate degree of emphasis.

This is done by varying the image size of the subject in relation to the total frame area, which normally means moving the camera in closer or further away (although, of course, a similar result can often be achieved by adjusting the focal length of the zoom lens as was discussed in the previous chapter).

There are three basic types of shot: Long Shot (LS), Medium Shot (MS) and Close-up (CU). These are sometimes further qualified to indicate their degree of 'longness' or 'closeness' as with *Extreme Long Shot* or *Big Close-up*. It should, perhaps, be made clear that it's the image size when considered in relation to the size of the original subject that really defines the type of shot rather than the actual camera-to-subject distance (or lens setting). For example, if we filled the picture with the head of a guinea pig, the animal would have been recorded only inches from the camera, whereas a similar shot of a hippopotamus would have needed a much greater camera-to-subject distance, yet both shots would be called close-ups.

In view of this 'relativity factor', it's probably simplest to talk about these different types of shots in terms of the human figure.

## Long shot

This, you might say, is the foundation stone of movie making. It's a general purpose kind of shot that's used for establishing a location at the start of a sequence, or the spacial relationship of characters to each other or to objects within the setting. It's the shot that gives viewers their bearings before we start involving them in the action by moving in closer.

In a long shot, the whole of a human figure would be shown. The character might be little more than a tiny silhouette against the sky on the horizon, in which case it would be the extreme LS mentioned above. On the other hand, the character might be much nearer to the camera — say, a girl walking along a country lane — so that her head and feet are just within the borders of the frame; this would usually be called a medium LS. But in either of these two cases, the intention would probably have been to establish the character in a particular location which is one of the LS's main functions.

The LS, by its very nature, generally includes a number of elements in the picture at the same time, so it's important that some care is taken with the composition (see Chapter 9) so that the eye is led naturally to the main subject.

A frequent mistake of beginners in extreme

*EXTREME LONG SHOT — shows wide expanse of location with a human figure as, perhaps, little more than a speck on the skyline. Useful for suggesting isolation or loneliness.*

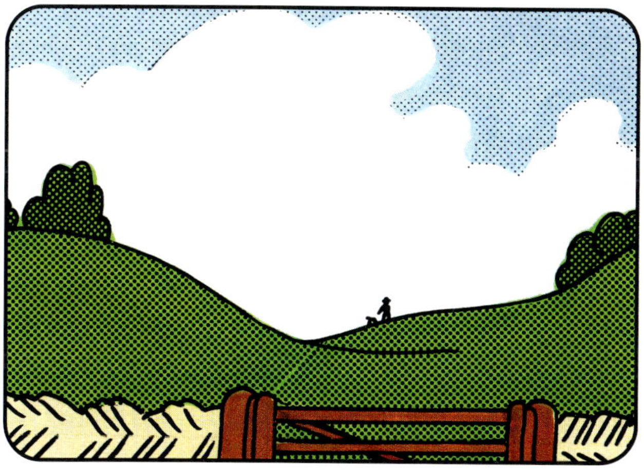

*LONG SHOT — allows us a closer look at the character, but still only gives a general impression of what he's like.*

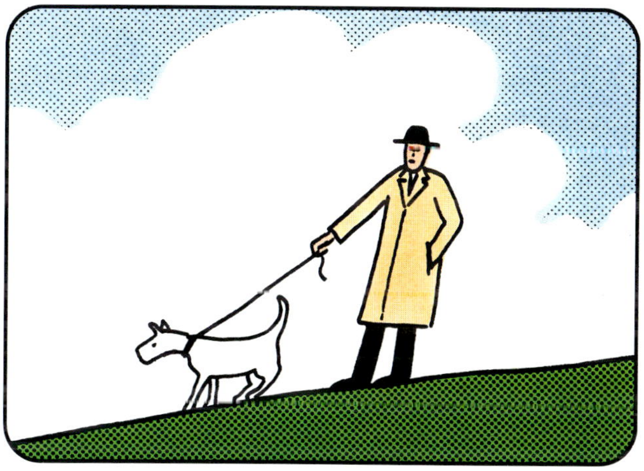

*MEDIUM SHOT — Now we can really see who the character is and observe all but the subtlest changes in his expression.*

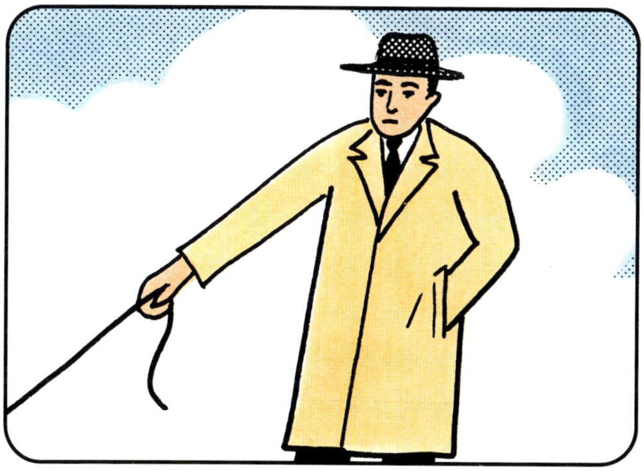

# 5

*CLOSE-UP — gives us the feeling we're meeting the character face to face and getting to know him personally.*

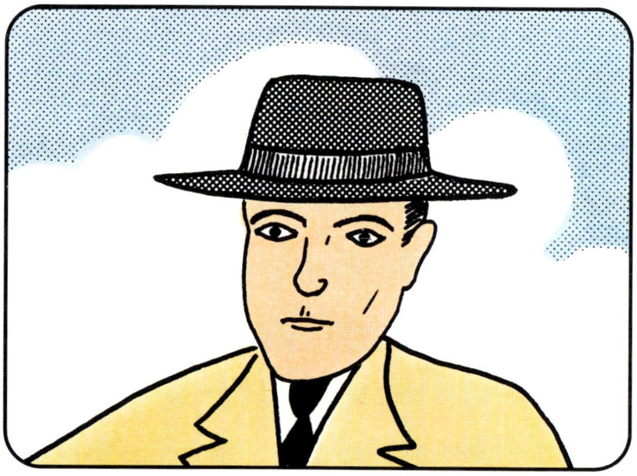

*BIG CLOSE-UP — A powerful image to use if the character starts to dominate. Also useful, when expression is reflective, to lead into a flash-back.*

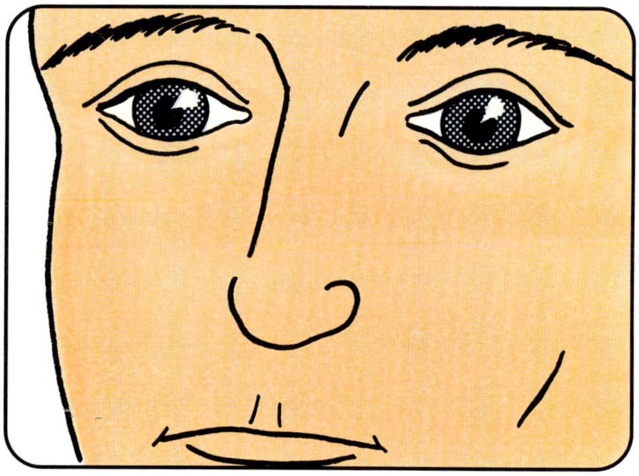

long shots — especially seascapes — is to have the skyline exactly half-way up the frame; this is always weak because it causes the picture to 'fall in half' through being divided across its centre. A far stronger and more pleasing composition is achieved by placing the horizon somewhere between a quarter and a third of the way up the frame.

An exception to this rule would be if the shot were being taken from an elevated viewpoint, as when looking out to sea from the top of high cliffs. Here it would be best to set the skyline about two-thirds of the way up the frame; this would include more sea than sky in the picture and so strengthen the impression that we are looking down from a great height.

And while we're talking about skylines — always make sure you're holding the camera level so that you don't get the drunken 'lopsided horizon' effect.

## Medium Shot

This one — sometimes also called a Mid-Shot — is rather more difficult to define precisely because it's very much a matter of opinion as to where a medium long shot ends and a long medium shot begins. Similarly, you could also argue about close medium shots and medium close-ups.

In general, however, medium shot implies that the camera is set fairly near to the subject and includes something less than the full standing figure — say, from just above the knees up. It's probably the most used type of shot and comes into its own with interior sequences showing a couple of characters talking — perhaps sitting opposite each other at a table or side by side on a settee (this is often known as a *two-shot*).

Its virtue is that it gives the viewer a close enough look at the figures to see their expressions, while also showing something of their immediate surroundings. Next time you watch a situation comedy on TV, see how frequently it's used in this context.

### Close-up

This is the movie medium's unique and most potent expressive device. Its great power derives from the fact that it enables the movie maker to draw the viewer's attention instantaneously to any detail of the scene or action — a facility that's equally useful in a drama, documentary or family video.

A close-up of a character would normally show the head and shoulders, and such a shot might be used when an actor is delivering some specially significant line of dialogue, an effect of emphasis being created by the large image of the speaker's face. Alternatively, a close-up is often used to show some subtlety of reaction to a line that another character has spoken.

If the camera is moved in even nearer to a person so that just a portion of the face is tightly framed — eyebrows to mouth, say — the result would be a big CU. Such an extreme image should be kept in reserve for those occasions when it's really needed for emphasis.

As I hope I've shown, close-ups are extremely useful shots, but for this very reason they need to be employed with discretion. Their impact can quickly be lost if they're peppered about a sequence indiscriminately without regard for their proper function.

Of course, there's nothing wrong with building an entire sequence from close-ups if it's done with some special purpose in mind. This might be to treat some piece of action impressionistically,

*Close-up is really a relative term, depending on the size of the subject. The shot of the hippo would obviously need to be taken at a much greater camera-to-subject distance than that used for the guinea pig, but both shots would be described as close-ups.*

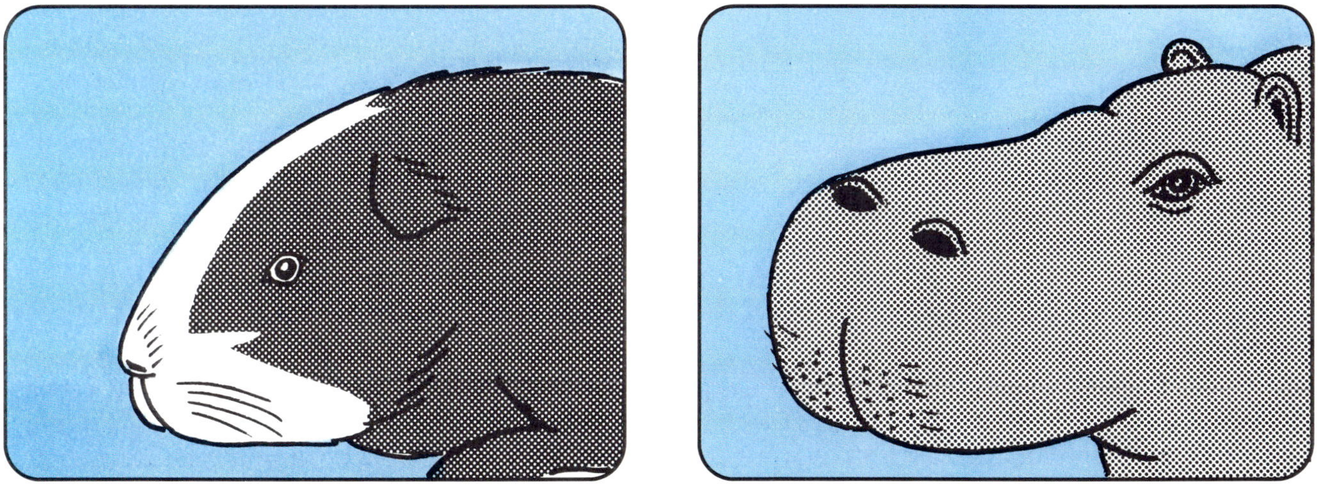

perhaps the preparations prior to setting off on holiday — suitcases being got out and dusted off, clothes being folded and packed, drawers of a desk being searched for elusive passports, and so on.

And it hardly needs saying that bold close-ups are the shots to use for capturing expressions of children. It's obviously best to take these with the zoom set at telephoto so that the young performers aren't inhibited or distracted by your sticking the camera lens right in their faces.

### Tempo and image size

The tempo or pace of a sequence will, to some extent, be determined by the types of shots that compose it. The long shot, as I've mentioned, is often used to establish a location, which means that it must be allowed to remain on the screen long enough for viewers to take in all the essential details.

When it comes to medium shots, there isn't really any hard and fast rule. Being such an all-purpose kind of shot the MS can feature in a quickly-cut sequence or take its place in a more leisurely episode and be held on the screen for as long as anything interesting is happening.

The boldness of an image in close-up instantly registers with the viewer and so can be whisked on and off the screen in a flash — a technique that's often employed in television commercials to put over the maximum amount of message in the shortest possible time.

The point I've been trying to put across in this chapter is that there's a type of shot suited to the needs of any particular movie situation, and it's really not difficult to decide which is right for the job. It all boils down to one simple question: what do we want the viewer to *see* at a given moment in our movie?

If we just stop to think and answer that question objectively, we shall have no problem making up our minds about how large the subject should loom on the screen, and planning a sequence in terms of long, medium and close shots will soon become second nature.

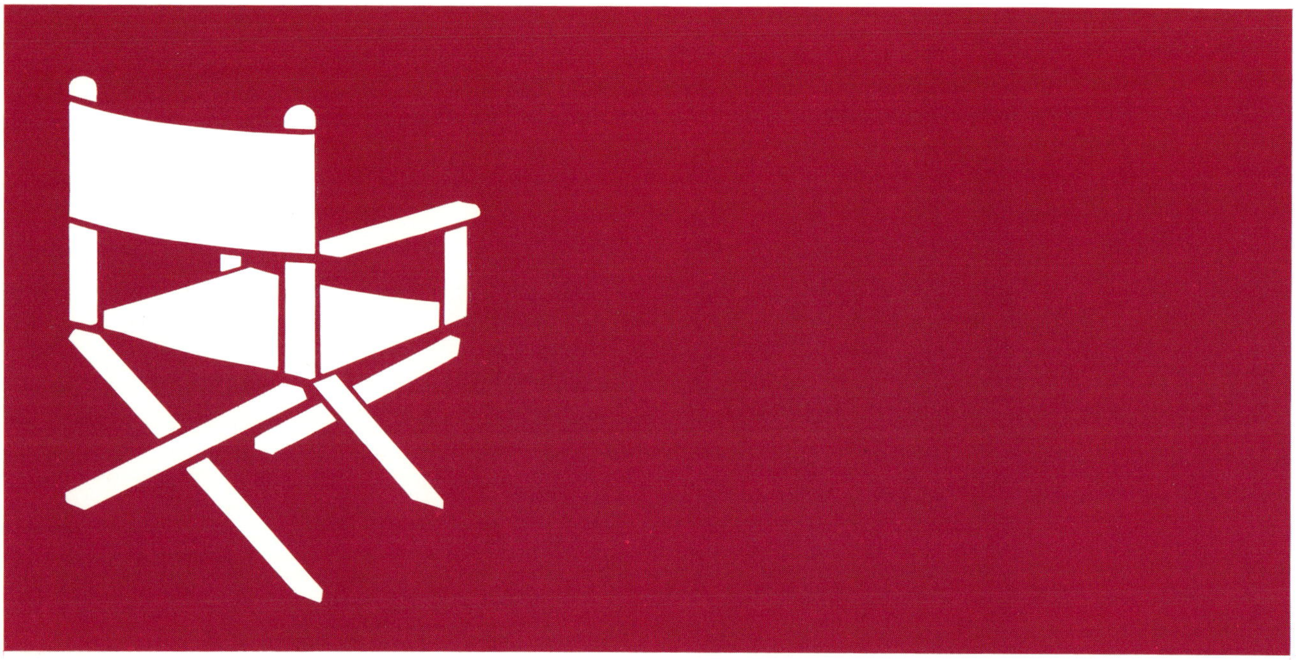

## Chapter 6
# Camera Movement

The trouble is that moving the camera is just *too* easy; all we have to do is swing it from side to side, up and down and round about as the fancy takes us. The consequence is that the beginner tends to abuse the technique unmercifully, unless he or she has been sternly warned against it.

But why is this? Well, I think it's largely because when newcomers to video first look through a camera viewfinder, they are conscious that, compared to still cameras, the area taken in seems rather limited — at least, when the lens is at its 'normal' zoom setting. They feel that they can't 'get everything in' and so start waving the camera all over the scene to try and remedy what seems to be a shortcoming. What they don't realise is the distressing effect this undisciplined camera movement — which has been given the highly appropriate name of *hosepiping* — will have upon their future viewers. When played back, the picture will skid drunkenly from side to side, with the result that those watching can see nothing properly and long for some seasickness pills.

So the golden rule for beginners is: *keep the*

# 6

*camera still and let the subject do the moving.* If you have the feeling you aren't 'getting everything in', then use separate shots giving two or more views of the subject, or possibly adjust the zoom to a wider angle or simply move the camera further back if this is practical.

Now you may be thinking that this advice to keep the camera still is rather unenterprising in a chapter headed 'Camera Movement'. But as I think you will readily see, we've got to learn to keep it still before we can move it effectively. Also, there is all the difference in the world between haphazard, unplanned movement, and moves that have been carefully thought out and used for some practical or dramatic effect. Such purposeful movement is an important part of movie technique, and the various forms it can take is what we're going to consider in detail now.

## Panning

This is the most basic type of camera movement, and is achieved by swivelling the camera to the left or right about its vertical axis. (In practice, if the camera is hand-held, it's more likely to be swung around the operator's vertical axis, but this doesn't make any significant difference.)

The effect of a panning shot on screen is that the scene appears to move to the left or right in the opposite direction to the original camera movement. A moment's thought will show why this is so: the screen, unlike the frame of the viewfinder, is fixed in space so all movement is seen relative to this static 'window' on the scene.

There are two main reasons for using a panning shot, the most useful and legitimate being to follow a moving subject that would otherwise pass out of frame too quickly. As well as having this purely utilitarian function, the technique can have the effect of emphasising the speed of a rapidly moving subject. If, for example, we were to set the zoom lens at telephoto and follow the action of a galloping horse, the animal itself could be held sharply in frame while the unimportant background detail was reduced to a streaking blur.

A point to bear in mind when panning with a moving subject in this way is always to leave more space in front of the subject than behind it when framing the shot. So doing will avoid the impression that the horse is gaining on the camera operator and is about to bolt out of the picture.

It's worth mentioning, in connection with this type of panning shot, that if the subject is following a circular route and the camera is set up at the centre of the circle, the effect is very similar to a lateral tracking shot (see later) and is far easier to contrive.

The other purpose of panning is to include in one continuous shot subject matter that is too widely spaced to be covered by a static composition. As I've already suggested, in such a situation it may be preferable to take two or three separate shots instead, but there are nevertheless occasions when it's perfectly legitimate, not to say dramatically more effective, to cover them with a pan. Such a case would be if we wanted to establish clearly the geographical relationship between two or more objects or characters in the location.

When panning to include an extended subject, such as a landscape, the camera movement should always be slow and steady so that the viewers can really see the details of the scene passing before them. Another point to remember is that the shot should begin and end with the camera *at rest* and lined up on some feature of interest in such a way that a pleasing composition is achieved. This is something that's unlikely to happen by chance, so the shot should always be rehearsed — perhaps a couple of times — before the recording is acually made.

There's one other type of pan that goes against everything I've advised in the previous paragraph — but for a very specific reason. This is the *zip* (or *whip*) pan that can be used as a transition device between two (possibly related) sequences. It works like this.

At the end of the final shot of the outgoing sequence, the camera is swung swiftly to the right so that the image on the screen becomes a blur. The next sequence is begun by swinging the camera, again to the right, then bringing it to rest on the new scene so that the image resolves out of a blur. Later, the end and beginning of the two shots can be edited (see Chapter 11) so that the blurred sections are reduced to a total of about half a second. The result when replayed will give the illusion that the camera is panning rapidly from one scene to another. It's a handy device to use in a magazine-type video, for example, where you want to link the items in a punchy way.

Incidentally, if you find it difficult to end the zip pan of the second shot with the camera properly lined up on the subject, you can actually cut the blurred section of the first shot straight on to a

*A panning shot can be used to follow a moving subject. If it's travelling quickly and you use the telephoto setting of the zoom, the impression of speed is enhanced because the background is reduced to a streaking blur while the subject itself remains sharply defined. But remember, when using this technique, to leave more space in front of the subject than behind it — you don't want it to look as if the horse is about to bolt out of the picture!*

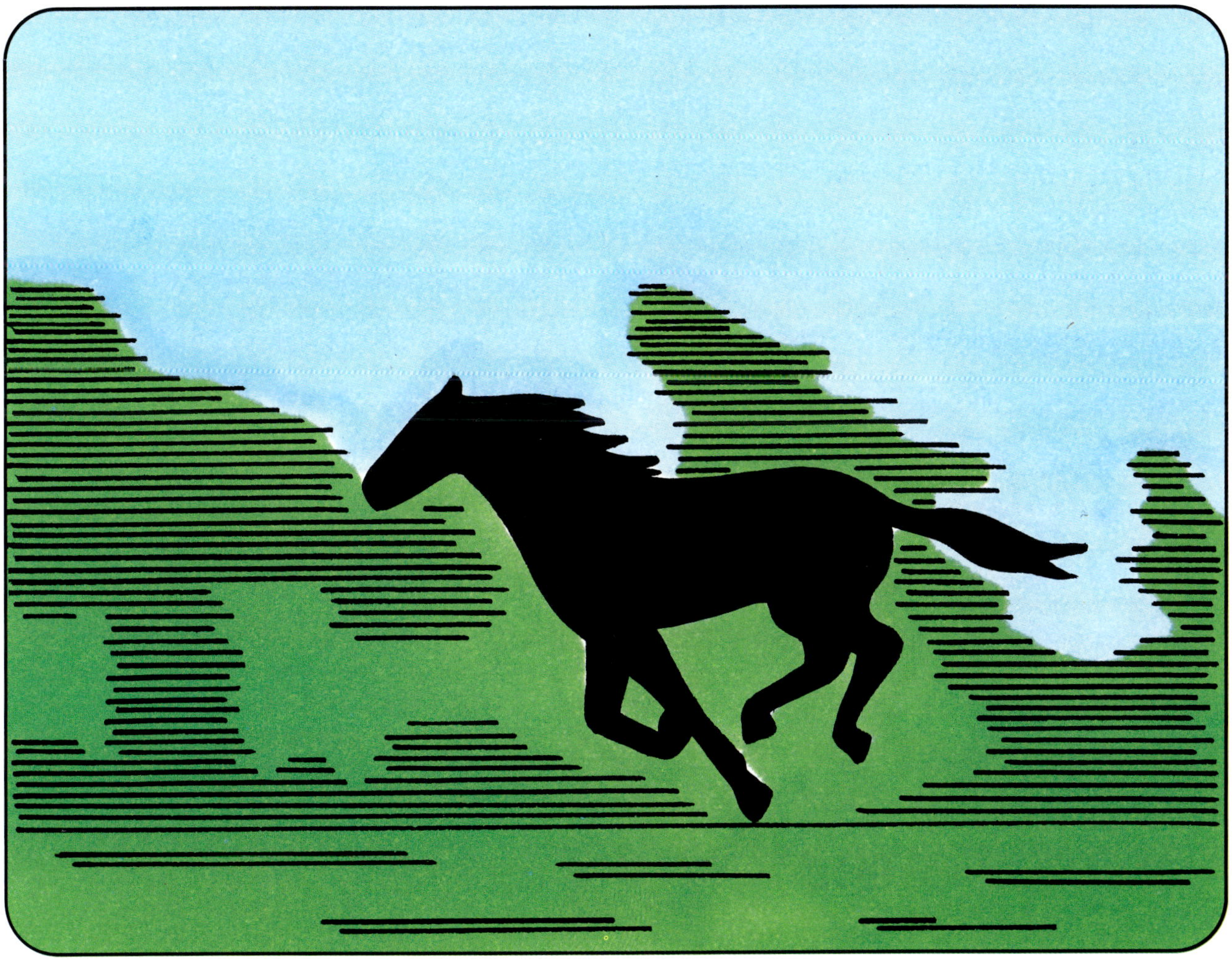

carefully composed static view of the new scene if you like. Thanks to the eye and brain's capacity for being deceived, the effect works almost equally well.

## Tilting

A tilt shot is realy nothing more than panning in a vertical direction. We might, for example, have a medium shot of a Spanish guitarist singing a serenade and then tilt the camera up to show his lady love on the balcony above. A tilt could also be used to emphasise the height of a very tall building.

A point to mention here is that such a shot might result in a lot of bright light from the sky flooding into the lens, so causing the automatic exposure system to react violently and close the aperture right down (see Chapter 2) to produce under-exposure of the main subject. It's best to guard against this by locking the meter on its original setting if your camera has this facility. Also we need to be careful never to turn our cameras directly into the sun when tilting, as permanent damage may be done to the image sensor.

I've talked about pans and tilts as though they are quite separate types of camera movement, but obviously the two can be combined so that the field of view may travel at any angle from one point of interest to another.

## Tracking

Tracking shots (or trucking shots as they're sometimes called) are rather more complicated for the amateur because they involve moving the camera and its operator backwards, forwards or sideways, or maybe on an even more complicated path. This naturally means mounting them on some sort of mobile base which can be moved easily and smoothly in the required direction(s).

Professionals make use of wheeled platforms called *dollies* or more elaborate *cranes* when a vertical movement is also involved, but usually the amateur has to be content with rather

*If you need to establish the geographical relationship of two features of a scene that can't be included in the same static shot, the answer is to pan the camera smoothly from one to the other. To achieve an effective composition of both the boat and the lighthouse, when the camera is stationary at the beginning and end of the shot, it's necessary here to pan at a slight angle to the horizontal. To make sure you get the movement and positioning right, it's wise to rehearse the pan a couple of times before actually taking the shot.*

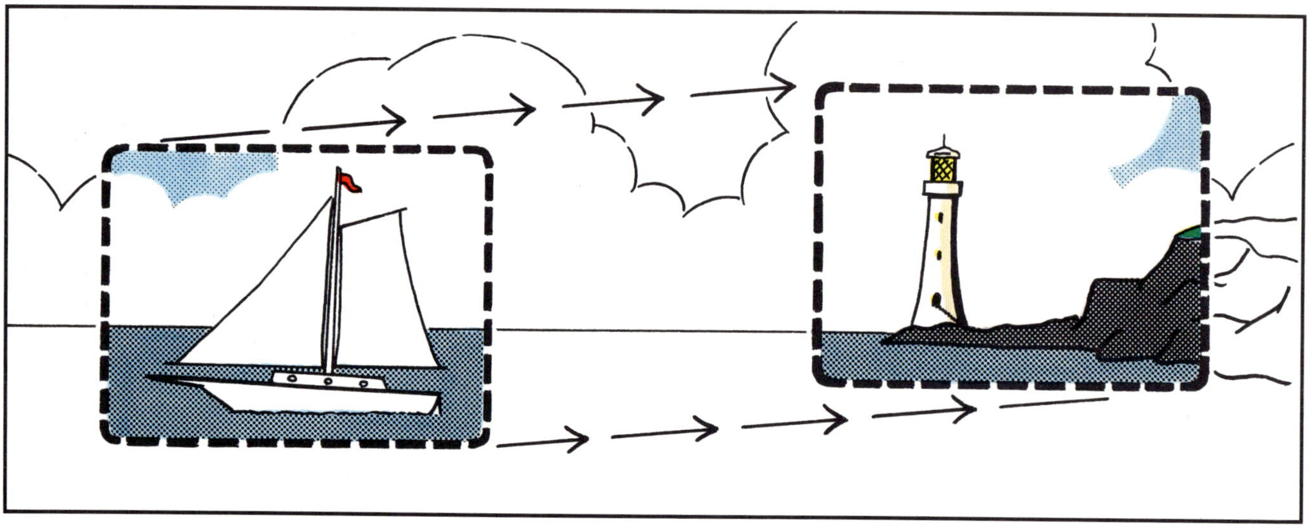

humbler improvised equipment. Wheelchairs, old prams and tea trolleys are often pressed into service here, and where a rising movement has been called for, I've heard of fork-lift trucks being used with a fair measure of success — but obviously you should be especially careful if involved with this sort of potentially dangerous equipment.

From what I've said, you'll have realised that tracking is the most difficult type of camera movement, and for this reason it tends to be

*The top diagram shows how the professionals take a lateral tracking shot by mounting the camera, A, on a wheeled platform that travels on rails along a course parallel to the moving subject — in this case a man walking. The lower drawing illustrates how amateurs, without such sophisticated gear, can achieve a very similar effect by having the subject proceed in a semi-circle. The camera, B, is placed at the circle's centre and panned to keep the man in frame. Note that for the purpose of illustration, the scale of this sketch has been compressed; in practice the circle would usually have a greater radius with the camera set up correspondingly further away from the subject.*

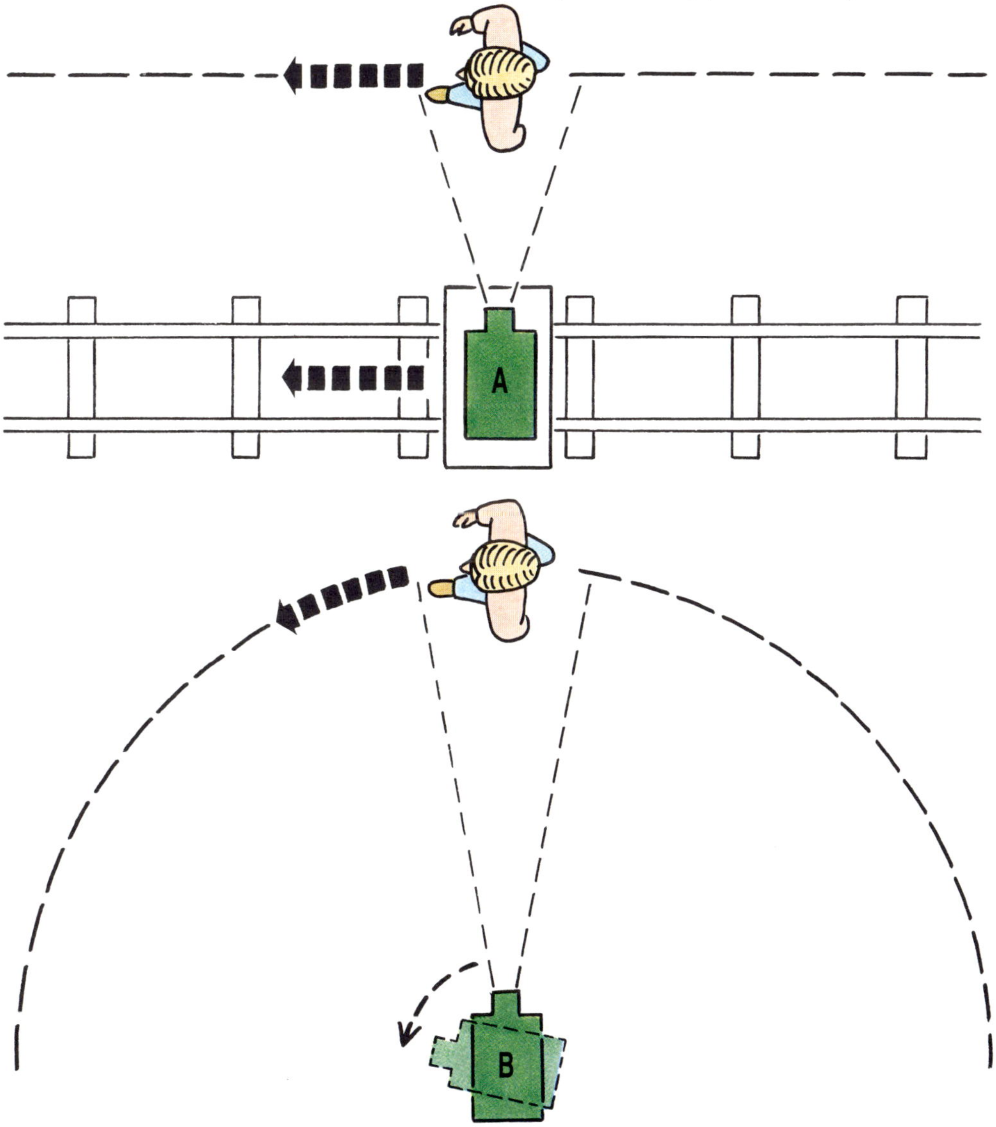

restricted (at least in amateur circles) to those productions where a shot can be set up and rehearsed to ensure that everything will, literally, run smoothly.

There are three basic types of tracking shot.

The *track-in* (moving the camera towards the subject) is frequently used in a dramatic context to concentrate the viewer's attention on some significant detail of the action. It's also very useful for giving a feeling of depth to a scene — especially if there are objects or people on either side of the camera — because of the changing perspective effect. This is something that you don't experience with a zoom shot, as was explained in Chapter 4.

The *track-back* (or *track-out)* is the opposite to a track-in and can, of course, be used for this depth-creating effect. However, it's often employed to put a character or object in context by gradually revealing the surrounding scene. This is sometimes done to play a trick on the viewer by suggesting one setting in the close shot and showing it to be something quite different as the camera tracks back.

Another purpose of the tracking shot is to keep a moving subject in the picture, which can be done by travelling the camera along a parallel course. In practice, the camera mount usually runs a little behind or in front of the subject (according to the needs of the action) to give an oblique rather than a direct broadside view. I mention this type of shot for the sake of completeness, because it's really beyond the resources of most amateurs, requiring as it usually does a dolly running on a specially laid track.

You can, however, sometimes manage to obtain successful tracking shots of this type by shooting from a car. Obviously the ground must not be too rough, and it's a good plan to let some of the air out of the tyres as this helps to minimise any bumps (but remember to restore them to normal pressure afterwards!). Using a wide-angle setting of the zoom will also enhance the feeling of a smoother progress.

Camera movements are particularly useful in those video making situations where long takes are likely to be the order of the day. They can be very helpful when shooting interview and group discussion programmes with a single-camera set-up where you may need to pan from one speaker to another so that the conversation can proceed without interruption. But the cardinal rule still applies that any movement must be *purposeful.* As I advised when talking earlier about zooming: if you're in any doubt about whether to move the camera — then don't.

Chapter 7

# Continuity: Keeping a Sense of Direction

It's often been pointed out that movie technique reflects the way we observe a situation in real life. Our eyes flit from one detail of the scene to another, enabling us to build up an overall mental picture of what is going on.

But there is one very important difference: our real-life observation is done from a subjective viewpoint of which we are continuously aware — in other words, we are always conscious of our spacial relationship to the elements of the subject our attention is focused on at any given moment. This feeling of instant orientation is lacking in the movie watching experience, so it's the business of movie makers to make sure they do nothing to cause viewers to lose their bearings.

Obviously, as I mentioned in Chapter 5, a long shot can be a great help when it comes to establishing initially the general layout of a particular setting. But once we start coming in closer for the detail shots that will make up the major part of the sequence, we need to consider

# 7

every camera angle with great care to make sure we don't confuse or destroy the viewer's sense of direction.

There are two basic aspects of this question of screen direction, so we'll consider each individually.

### Direction of movement

Because the field of view covered by a video lens is relatively restricted, a travelling subject soon passes out of the frame if it's fairly near the camera. (Sometimes, as we saw in Chapter 6, the camera will pan or track to keep the subject in view, but this is a special case.) So to ensure that a consistent screen direction is maintained, if the subject leaves the picture on the *left*, it must re-enter from the *right* in the next shot, and vice versa.

Similarly, if two separate subjects are supposed to be travelling in the same direction — as when one person is following another — they must always be shown moving across the screen in the same direction. If this rule is not followed the impression will be created that they are moving towards each other.

*When a subject moves out of shot, it must always re-enter from the opposite side of the frame in the next. The rule still applies when a subject is moving towards and then away from the camera. Here the approaching man passes the camera on its right, so on screen is seen to exit frame right. When the camera is turned for the reverse-angle shot he re-appears, logically, from the left. To avoid 'jumping' the character from one side of the screen to the other, it's best to let him move completely out of frame in the first shot and cut to the next when he is well into picture.*

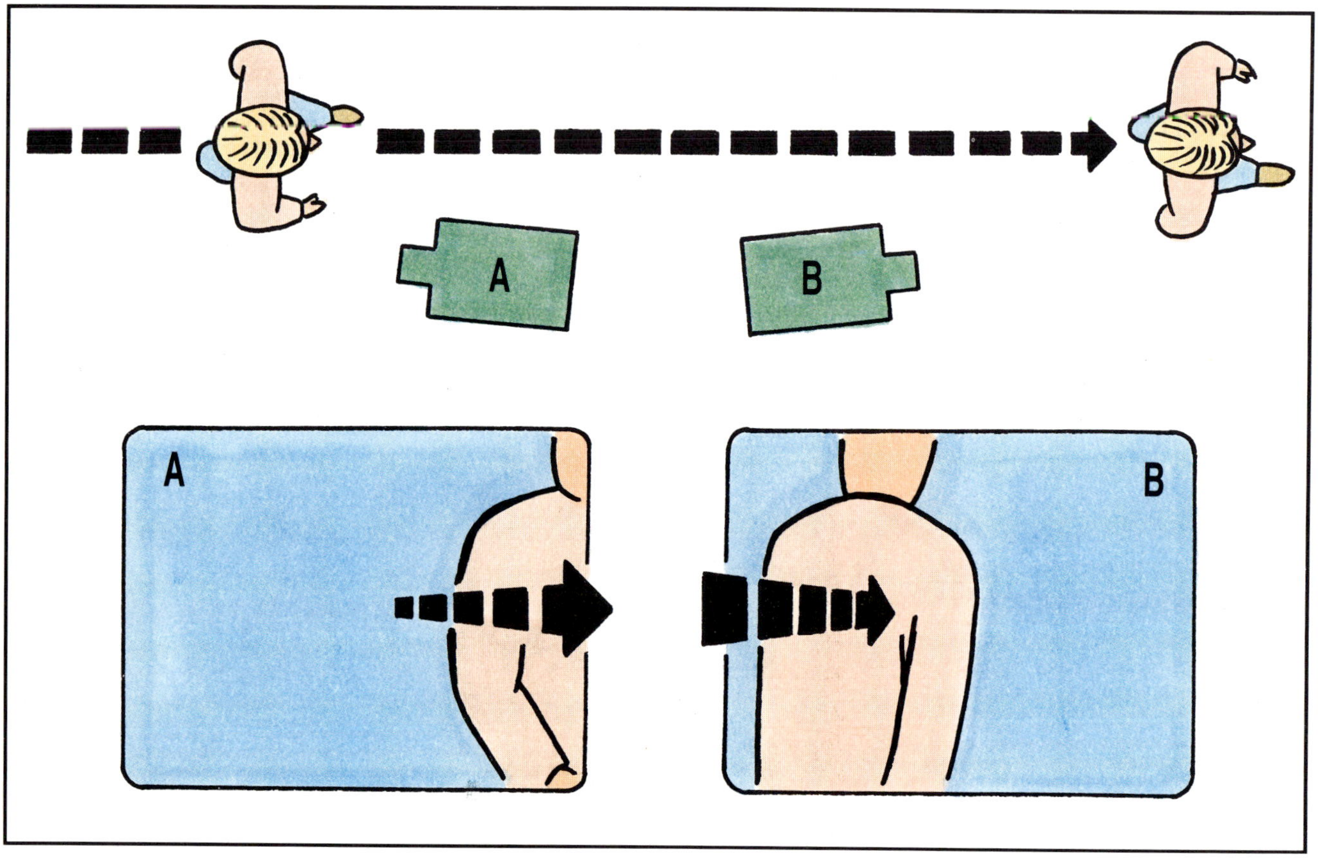

I might mention that consistency of direction is just as important if subjects are being held in frame with panning shots.

Now you may be thinking, rather impatiently: 'Yes, yes — of course I'd show the subject(s) travelling in the same screen direction from shot to shot. Why would anyone do otherwise?"

Why indeed? But it does happen — usually by accident. And the accident occurs because the camera has 'crossed the line'.

### The 180-Degree Rule

What line am I talking about? The one running along the path that the subject is following. We can move the camera to any position within an angle of 180 degrees on one side of this imaginary boundary, but step across to the other side and the direction of the subject on screen will immediately become reversed — the effect we want to avoid at all cost. Believe me, this can easily happen in the heat of the moment when shooting, especially if a sequence is being recorded off the cuff as opportunity offers. For this reason, a conscious effort must always be made to bear the 180-Degree Rule in mind.

If, after taking some vital unrepeatable shot, you find that you have inadvertently crossed the line, the situation can often be retrieved by use of a neutral direction shot. This is one where the subject comes straight towards, or recedes from the camera; if it's inserted between two shots showing movement in contrary directions, it will to a large extent mask the discrepancy. However, it's very much a 'salvage operation', and obviously it's better to take care so that you don't get things wrong in the first place.

Naturally, it's quite in order for screen direction to be changed if this is necessary to the situation — perhaps for some dramatic reason, characters have to be shown retracing their steps. But in such a case they must *be seen* to make the change, perhaps by making a U-turn or it could be established that they've been taking a circular route.

This principle of screen direction is easy enough to grasp when the subject is travelling more or less broadside on to the camera, but confusion can sometimes arise when movement is almost along the axis of the lens. Suppose we have a man walking down a path towards the camera, leaving picture right of frame, re-entering in a reverse angle and striding off into the distance. Which side of the screen should he enter in the reverse angle?

The answer, of course, is from the left — all quite logical when you think about it. Obviously the walker can only pass on one side of the camera, which is another way of saying that the 180-Degree Rule has been obeyed; the fact that the camera is now facing in the opposite direction (for the reverse angle) doesn't affect the issue. This means that as the man exits right, he will automatically re-enter left in the following shot.

If you're still not convinced, think of it this way. Instead of making two separate shots, we could pan the camera rapidly around with the man as he passes. Here the camera has certainly not 'crossed the line' because its *position* is unchanged, and as it's using the same angles (albeit continuously) that gave us the exit-right-enter left result, these on-screen directions must be correct.

What worries some people about this is that at the cut there can be a rather disturbing 'jump' of the character from one side of the frame to the other. But the solution lies in editing (see Chapter 11): we should allow the walker to pass completely out of frame and then cut to the reverse angle with him already well into picture. So doing will diminish the apparent 'jump'.

I should, perhaps, make it clear that the 180-Degree Rule isn't something that only applies in a fictional context; it's equally important to observe it when making personal movies if these are to have a professional polish. If, for example, you were shooting a school cross-country run, the result would be total confusion if various groups of competitors weren't seen running consistently in the same direction. Likewise, if the family are shown engaged in some beach game, when the ball is thrown out of frame left in one shot, it must come in from the right in the next.

## Eyelines

The 180-Degree Rule is equally important when taking shots in which two or more characters are talking to each other. If it's not obeyed, the *eyelines* — the apparent directions in which they are looking — will be all wrong. It would obviously make nonsense of a scene where a couple are supposed to be having a head-on confrontation if, from the viewer's point of view, they seem to be facing in the same direction.

As with the action scenes we've already considered, the trick is once again to avoid crossing the line — in this case, one drawn through the centre of the two characters as they face each other. As before, we can move through any angle up to 180 degrees when adopting camera positions to favour first one speaker and then the other, but we mustn't cross that line. If we do, the character seen in that shot will immediately appear to be looking in the reverse (and so wrong) direction.

Shooting a scene with two people is fairly straightforward, but if a third is introduced, things get more complicated because the range of camera positions is further restricted.

Imagine a group of three women sitting in a triangle formation and chatting. Speaker A is at the apex of the triangle and speakers B and C are at either end of a line forming the base. Now if we are to ensure the eyelines of *all* speakers are *always* correct in relation to each other, we have a choice of *one* of three limited areas — we'll call them One, Two and Three — in which to set up the camera.

These areas are bordered by a side of the triangle and extensions of the adjacent sides, so, for example, Area One is bounded by the 'side' between speakers A and B, the eyeline between speakers A and C extended beyond A, and the eyeline between speakers B and C extended beyond B. If all this sounds a bit like a geometrical theorem gone mad, I'm sure a glance at the accompanying diagram (page 48) will make everything crystal clear.

Having decided which area we're going to shoot from, we must stay in it throughout the sequence, unless some action takes place that clearly indicates to the viewer that the viewpoint has logically changed — this might be a fourth character who's watching the group being seen to move to the other side of the room. However, if we keep within our chosen bounds, the eyelines of each speaker will be guaranteed correct whoever she's addressing and irrespective of whether one, two or all three speakers are in shot.

Let's assume we decide to shoot from Area Two (as illustrated). In this case Speaker A will always be looking to the left, whether she's addressing B or C, who will both, in their turn, be looking right when talking to A. If the camera is turned to favour B, she will still be looking to the right when talking to A and slightly to the left when addressing C. When the camera is turned on C, she will still be looking in the correct direction, right, to speak to A and also to the right when addressing B, which is logical in relation to B's leftward look at her.

So you see, it all works like a charm, *provided you stay put in Area Two.* The inexperienced video maker might be tempted into Area Three, bordered by the 'base' of the triangle (see diagram), to get a shot favouring A. This would immediately upset the eyeline applecart because A would now appear to be looking to the right when addressing C when previously she had been seen looking to the left. So I repeat: you *must* stay in your chosen area when shooting a three-way conversation, but within that area you can set up the camera anywhere you like, and shoot at any angle or distance in the certain knowledge that all the eyelines will be correct.

*When shooting a scene in which two characters talk or otherwise confront each other, it's essential that the camera stays on one side of the 180-degree line. If this rule is not obeyed, one of the characters will suddenly appear to be facing in the wrong direction, as can be seen in the shaded picture which was taken from the incorrect 'across the line' position at C.*

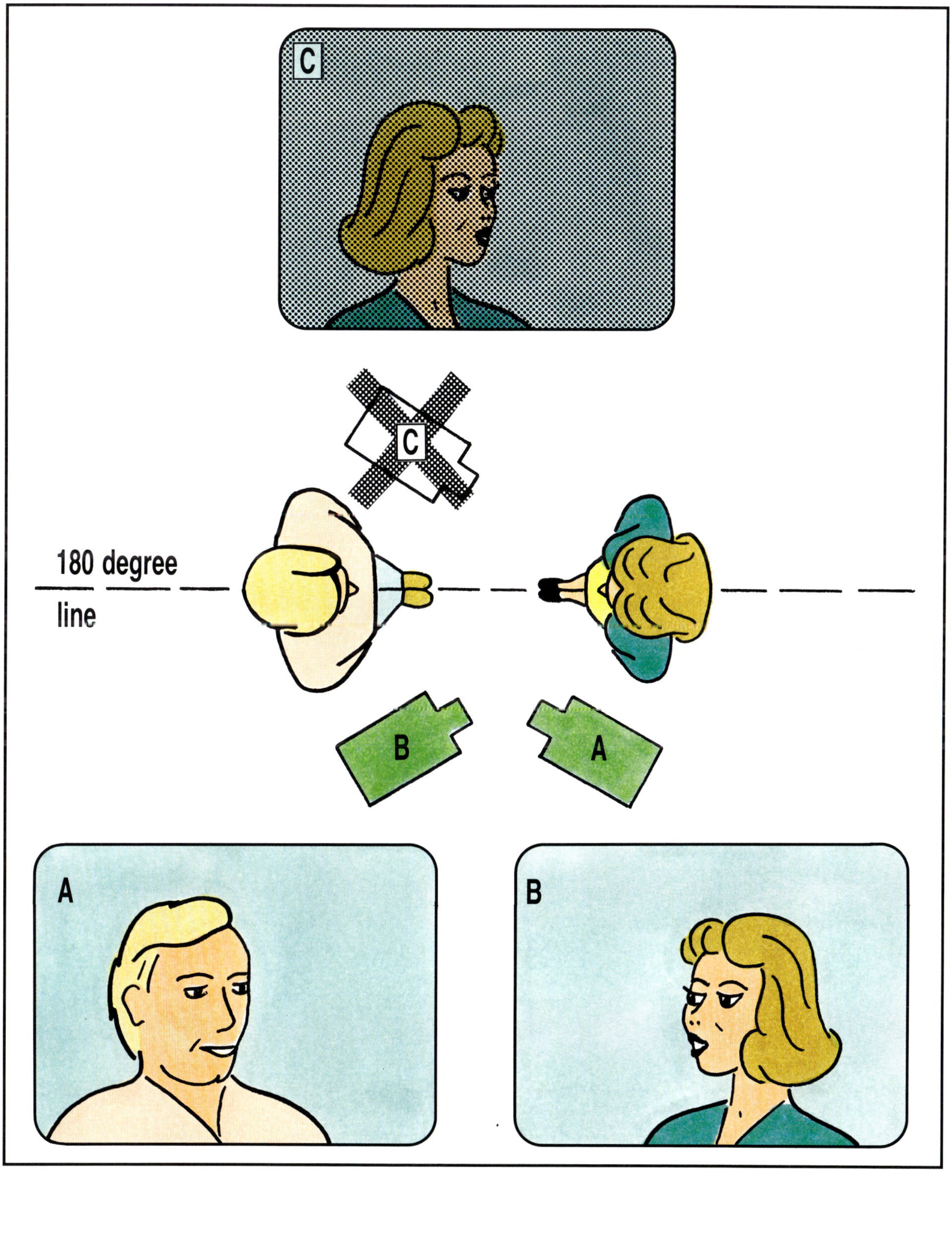

# 7

## 'They went that-a-way!'

There's sometimes confusion over what we might call the 'they went that-a-way' situation, or less whimsically, the spectator viewpoint. The scene is a race course. A long shot shows the horses flashing by, left to right, as seen from the grandstand. We now cut to a reverse angle medium shot of two characters turning their heads to look after the animals as they pass. The

*The 'threesome triangle'. When shooting this type of three-way conversation scene, the movie maker must decide to place the camera in just* one *of the areas bordered by a side of the triangle (in this case Area two has been chosen). It can then shoot at any angle or from any distance and the eyelines of the three characters will always be correctly related to each other. But if it moves to either of the other two areas, the direction in which one or other of the characters is facing will be reversed on screen and 'geography' of the scene will become confused.*

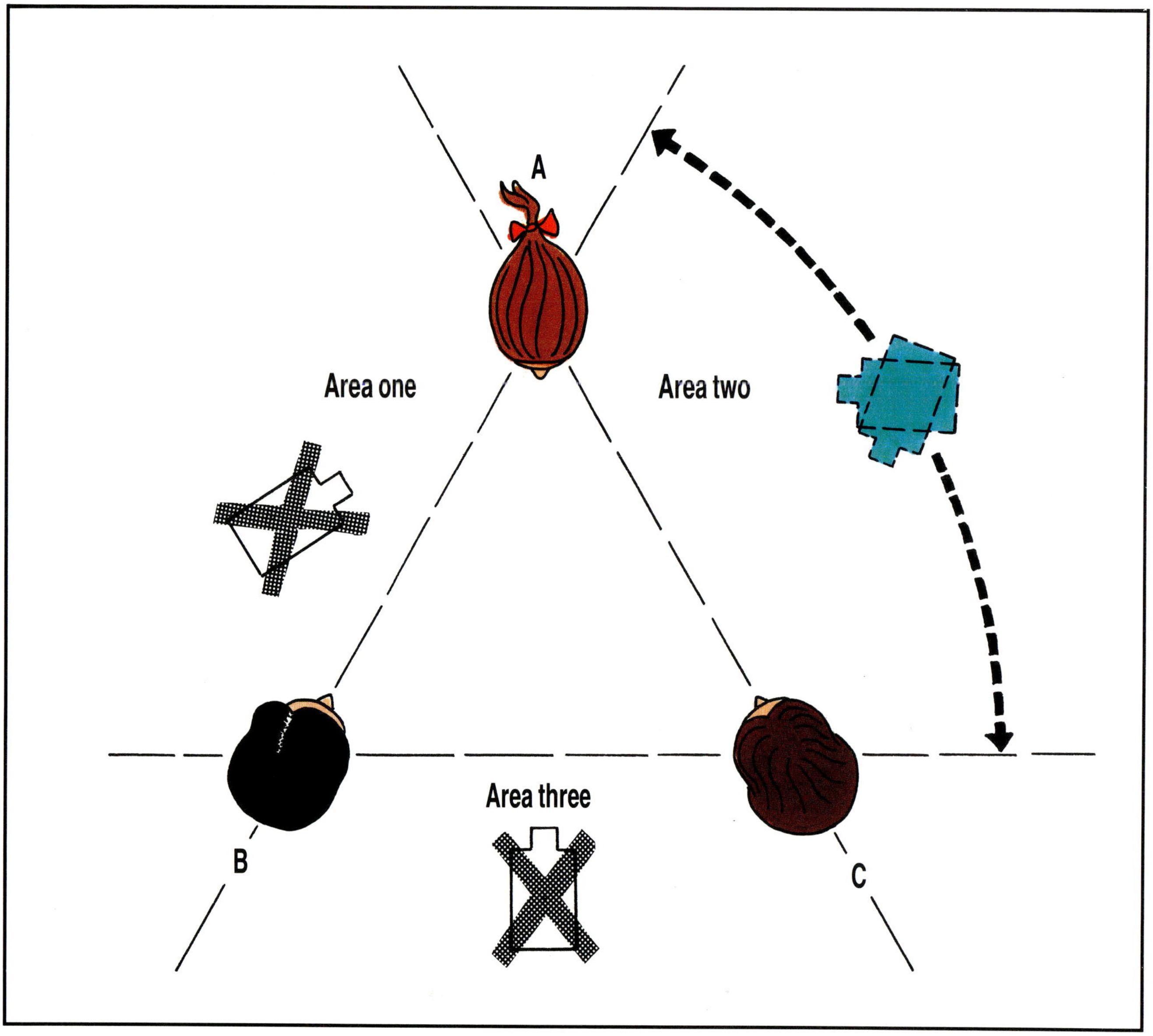

question is: which way should they be seen to look?

Some beginners might be tempted to think that because the previous shot has shown action moving across the screen from left to right, this is the direction in which the spectators should turn their heads. But remember that this is a *reverse angle*. Imagine the camera set up in front of the grandstand to take the reaction shot: the horses would be passing *behind it from right to left,* so, logically, if the spectators are following them, they must be seen looking in this direction, too.

Or take another case which could crop up in a travel video. A medium long shot shows a passenger looking out of a train travelling from right to left. If we want to follow with the passenger's view of the passing landscape; which way should the scenery move across the screen? The answer this time is *in the same direction as the train* in the preceding shot — that is, right to left. This is because the camera is now in the position of the passenger and is moving left to right relative to the scenery which thus passes across the screen in a contrary direction.

This question of the direction of 'moving' scenery brings to mind another point that often arises with travel movies. Obviously on any journey, features of interest are likely to appear on either side of the conveyance in which we're travelling, but crossing to the other side to shoot them will cause the scenery to start 'moving' in the opposite direction to that already established, so giving the unwanted impression that we're going back to where we started.

Fortunately there's a simple way around the problem. Before crossing to the other side of the vehicle, we take a close-up of our featured passenger looking at the scenery as it passes in the original direction. Before cutting the shot, we get the observer to turn his or her head towards the view on the other side. Now when we follow with a shot of the scenery passing in the reverse direction, it won't have the same jarring effect on viewers because they'll appreciate they're seeing things from the passenger's viewpoint which has been seen to change. As with so much in movie making, it's a matter of applied psychology.

# 7

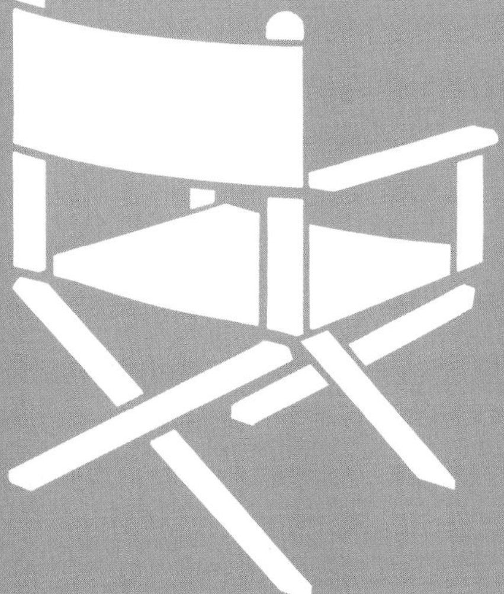

# Chapter 8
# Continuity: The Illusion of Non-Stop Action

If you were to ask me what I thought was the single most important element in successful movie making, I would unhestitatingly answer: achieving a feeling of logical progression from one shot to the next. This reply might surprise some people; surely this can't be as important as good photography? Oh, but it can, as I hope to demonstrate.

We've already discovered from previous chapters, and indeed from our every-day experience of films and television, that movies are made up of separate shots which, from a diversity of viewpoints, create a total impression of the action. But the thing is, it doesn't matter how individually brilliant those shots may be, if they aren't linked together by a feeling of visual logic the whole structure of the movie will collapse. Attempting to construct a video without this vital element is like trying to build a brick wall without cement.

# 8

Such shot-to-shot visual logic is really at the basis of what is called *continuity*. This term is sometimes used in the sense of simply ensuring that a girl is carrying the same handbag in consecutive shots that may have been taken days or even weeks apart, but it has much subtler implications than that. For one thing, it embraces all those psychological factors of cause-and-effect, action-and-reaction that I referred to in Chapter 1.

It also has to do with ensuring that physical action *appears* (and I emphasise that word for reasons that will become apparent in a minute) to flow smoothly and uninterrupted from one shot to the next. As this aspect is usually one of the major stumbling blocks for beginners, I'll deal with it first.

## Beware the jump-cut!

Somebody once said that drama is life with the dull bits left out, and the dictum applies equally (perhaps even more) to movies. However, we can't act on this principle simply by recording only the interesting bits and leaving it at that. If we did, and contented ourselves with switching the camera on and off according to whether there was anything exciting happening in front of it at a given moment, we'd produce a string of *jump-cuts* — the commonest and most-to-be-dreaded fault in amateur movies.

So that the nature of the problem and its various solutions will be quite clear, I'd better explain precisely what a jump-cut is. Putting it in a nutshell, a jump-cut occurs when the camcorder is stopped and then restarted on the same subject from the same viewpoint. When the result is played back, the viewer notices a very disturbing jolt where the action suddenly jumps forward because of the missing section that wasn't recorded during the time the camcorder was stopped.

There are two ways of dealing with jump-cuts: they can either be disguised by taking appropriate action at the time of shooting, or in some cases their unpleasant effect can be neutralised after the event at the editing stage by bridging the gap with another shot.

Let's take a specific example. Jump-cuts are at their most obvious in action where viewers are aware of some progressive development. Suppose we are shooting a scene of someone opening a parcel; this is not, perhaps, a very dramatic event in itself, but it may be necessary to the sequence in hand. The action could well take a minute or more in real time and to show it in its entirety would be boring. As explained above, to try and shorten the operation by just stopping and restarting the camcorder will produce jump-cuts. So what do we do? We simply remember the rule that most jump-cuts can be effectively masked by *changing the image size and the camera angle.*

We might begin our parcel opening with a medium shot of the package standing on a table; the opener takes up a pair of scissors, snips the string and starts to remove the wrapping. Recording could then be stopped while the bulk of the unwrapping process continues, to be restarted from another angle to give a closer view as the final covering is removed and the contents are revealed.

This is the simplest approach which works effectively in most cases and requires least thought and effort. However, it can be improved upon to make a flow of shots that's smoother and rather more dramatic. We could re-script the sequence like this:

MS — *Opener takes scissors and snips string of parcel.*
CU — *Opener's face wearing expression of anticipation.*
CU — *Hands removing wrapping.*
CU — *Opener's face* (angle as above) *now showing more excitement.*
MS — *The final wrapping is removed to reveal contents.*

This version is better than my first suggestion for a number of reasons. The most eagle-eyed viewer will not now notice any unnatural jumps in the action because all the apparent time is accounted for. The two occasions — and they need only be seconds — when the opener's face is seen are automatically accepted as allowing time for the stages of the unwrapping that aren't shown. This, of course, is quite illusory, but then illusion is what movies are all about.

In addition to having improved the sequence technically, we've also made it more dramatically satisfying by showing the reactions of the person involved, so adding human interest to a simple piece of business. Not only that, but we've applied the psychological 'cement' needed to bind the shots together by getting the viewer concerned about the outcome of the action. The facial close-ups of the opener raise a question —

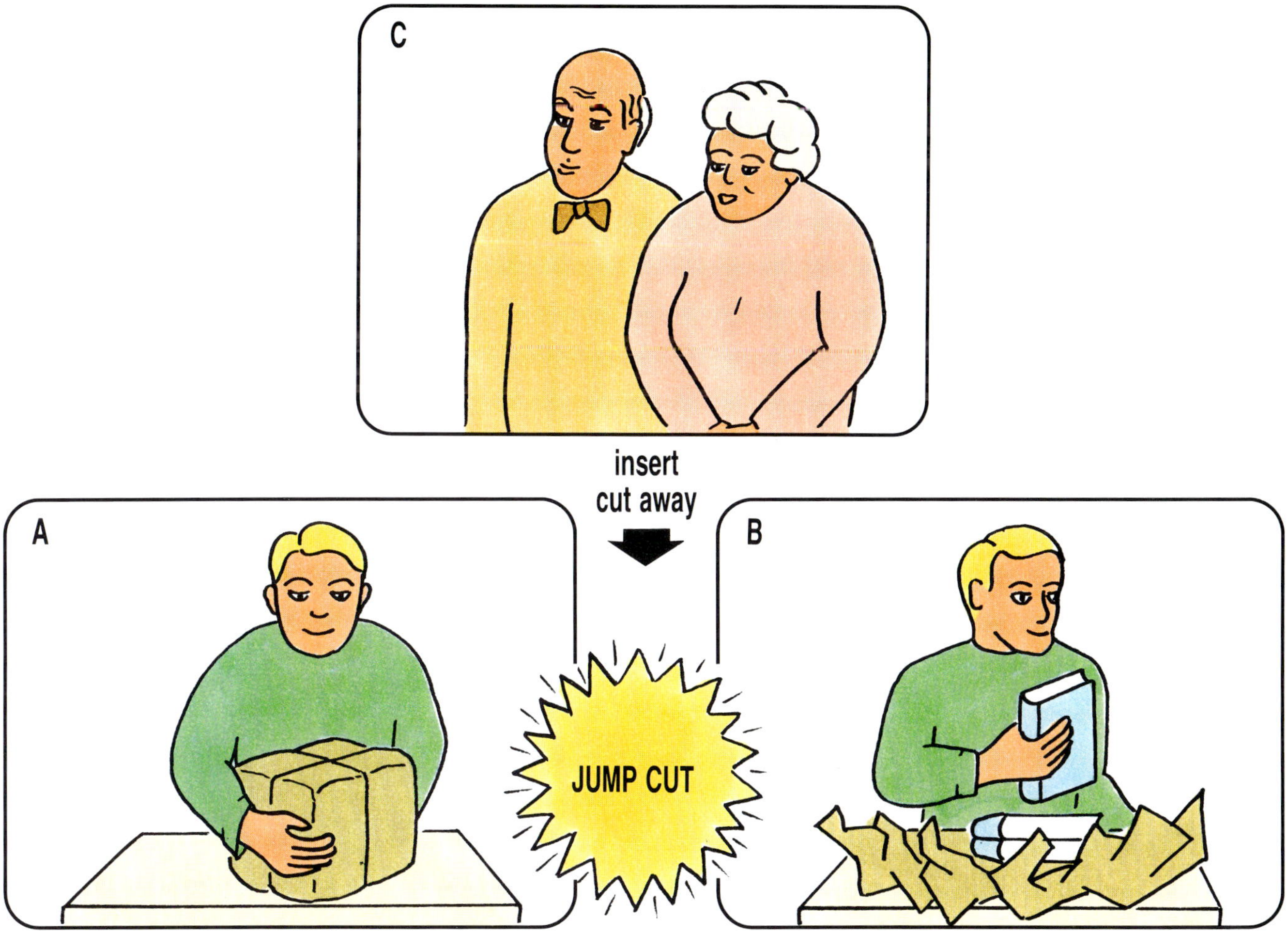


*Curing a jump-cut with a cut-away. Shot A was taken as the young man started to unwrap the parcel. Recording was then stopped and recommenced from the same angle to take shot B as he finished unwrapping and picked up the book. Result: a nasty jump-cut between the two shots. This can be put right simply by inserting cut-away shot C of the elderly couple watching as he unwraps the parcel. Note that the duration of shot C doesn't have to be anything like as long as the actual time it took to do the unwrapping.*

'I wonder what's in the parcel?" — which is then answered in the final shot showing the contents, so achieving that desirable feeling of logical progression.

### Cut-ins and cut-aways

Because the close-up of the hands is a detail of the main action it is called a *cut-in*. However, the facial close-ups are not an integral part of the action so they are called *cut-aways*. Ideally, a cut-away should be closely related to the main action, as here, but under the right circumstances it can be more remote. In this instance, if there were observers present, the parcel opening could be handled like this:

MS — *Opener takes scissors and snips string of parcel.*
MS — *Members of family watching opening business.*
MS — (original angle) *Final wrapping removed to reveal contents.*

If, in the context, the act of opening is not especially significant and you don't wish to dwell upon it, that might be the best way of staging the scene.

As well as bridging what would otherwise be a jump-cut, the cut-away can also serve the purpose of throwing some further light on the main action. Instead of the shot of the watching family in the previous example, we could insert a close-up of an array of birthday cards on a mantlepiece; this would tell viewers that the character they were watching was unwrapping a present.

But whichever of the foregoing methods we use, we have achieved the same objective: eliminating unimportant action without disrupting the visual continuity. In other words, we have telescoped time, a technique which can be used to condense any event to manageable proportions. And I think it's important to emphasis here that this needs to be done for artistic and not merely practical reasons. Just because video tape is relatively cheap, there's no virtue in recording the whole of some action if it has no intrinsic interest.

Although it has no direct bearing on my theme of continuity, as I've mentioned telescoping time

*If shot A is followed by shot B showing the completion of the unwrapping action from another angle and in a different image size, the unpleasant effect of the jump-cut is greatly reduced. However, the result is still not as smooth as that achieved with the cut-away technique.*

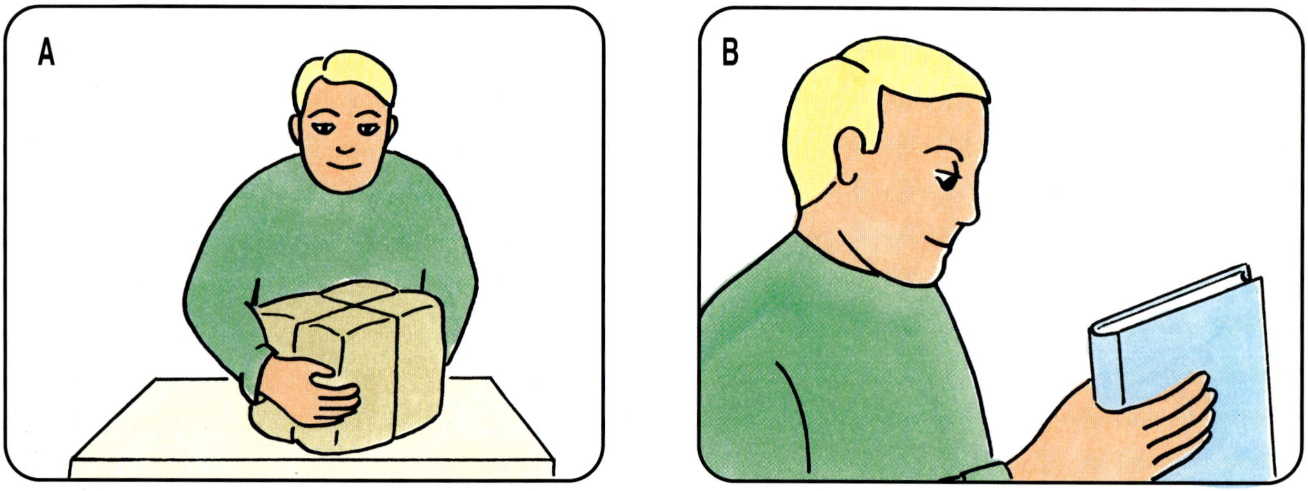

it's worth pointing out in passing that it can also be expanded. A cut-away or cut-in can obviously be used to increase the apparent length of time an action takes to perform. This is most likely to be useful in a dramatic context when we want to create suspense by prolonging an action.

Sometimes when professionals stage a spectacular crash, they will shoot this with a number of cameras from different angles, then at the editing stage overlap these various takes slightly by including in the incoming shot a second or so of action already covered in the outgoing shot so that the thrilling spectacle lasts longer on the screen than it took in real life. Cut-aways to spectator reaction can help here, too.

### Look and see

Another good example of the question-and-answer, action-and-reaction technique for binding shots together is what we might call the 'look and see' method. This depends on the fortunate psychological fact that when viewers see a character looking out of frame, they will automatically believe that the next shot shows what the person is looking at. This is particularly useful in the context of travelogue-type videos where we want to include plenty of scenic material yet maintain a feeling of progressive continuity and an element of human interest.

Perhaps you've been shooting the family in one setting and wish to change the location without too much of a visual jolt or having to resort to uninteresting 'walking about' shots. You could do it simply like this. Adopting a below-eye-level camera angle, compose the shot so that nothing is seen except a neutral sky background. Hold this for a second (literally) and then have one of your established characters walk into frame so that you get a close-up of him or her obviously looking at something off-screen.

Now cut to the view he of she is presumed to be looking at, and follow this with another close-up (it's best to use the same angle as the original shot) showing the reaction to what is seen. It's a simple device but has the two important virtues of giving the shots involved the essential feeling of logical progression, making viewers feel that the video is moving on, and allowing them to see some faces as well as places — always highly desirable in movies of this type.

### Cutting on action

I think few experienced movie makers would dispute that the smoothest and most satisfying progression from one shot to another is by the *cut on action*. This is used as a regular device by professionals to give viewers the illusion that although a sequence is composed of many individual shots they are watching an unbroken flow of action.

It tends to be employed when characters sit down or stand up, raise drinks to their lips, put on hats — any action, in fact, that can be made to 'flow' across the cut. One of its most frequent uses is to move unobtrusively from long shot into close-up. Suppose we have a cafe scene: a girl is sitting at a table with a cup of coffee in front of her; she seems unhappy and we want to give the viewer a closer look at her woebegone expression. One way might be to zoom slowly in on her face, but an equally valid method would be to switch to a close-up, cutting on action as she raises the coffee cup to take a sip.

As cuts on action depend on our having overlapping coverage at the end of the outgoing shot and the beginning of the new one, the business of the girl raising the cup to her lips has to be shot twice. So in the original medium shot the recording would continue until she was actually drinking, and the close-up would be started with the cup back on its saucer then coming into frame as it moves to her lips.

During editing, the cut would be made at an appropriate point in the overlap, which may not necessarily be at the mid point of the action. It's risky to lay down hard and fast rules on a subject where subtle differences in circumstance can alter a case, but as a general principle it's best to cut at the point where the action is about to be *concluded in the close-up*. So in our coffee-sipping example we would cut from medium shot to close-up when the cup has travelled about three-quarters of its journey. This creates the psychologically satisfying effect of introducing the new viewpoint at the completion of the action.

On the other hand, if a seated character has been shown in close-up and we want to cut on action to a medium shot as he gets up, the cut should be made just after he has started to move in the close-up, the bulk of the action being completed in the new shot. The reason for this is not difficult to see; there's no point in holding the uninteresting blurred image as the top half of the character's body moves upward in the frame.

From what I've been saying about cutting on action, it'll be apparent that its success depends on our being able to view the end and beginning of the two shots concerned and select the precise moments at which to make the edit — sometimes rather difficult for users of present domestic video equipment. It can just about be done by trial and error (see Chapter 11), but it's much simpler if the video maker has access to the more professional editing suites available at facility houses. However, it has to be said that if one can't manage cuts on action with real precision, it's best not to attempt them because nothing looks worse than when they're badly done.

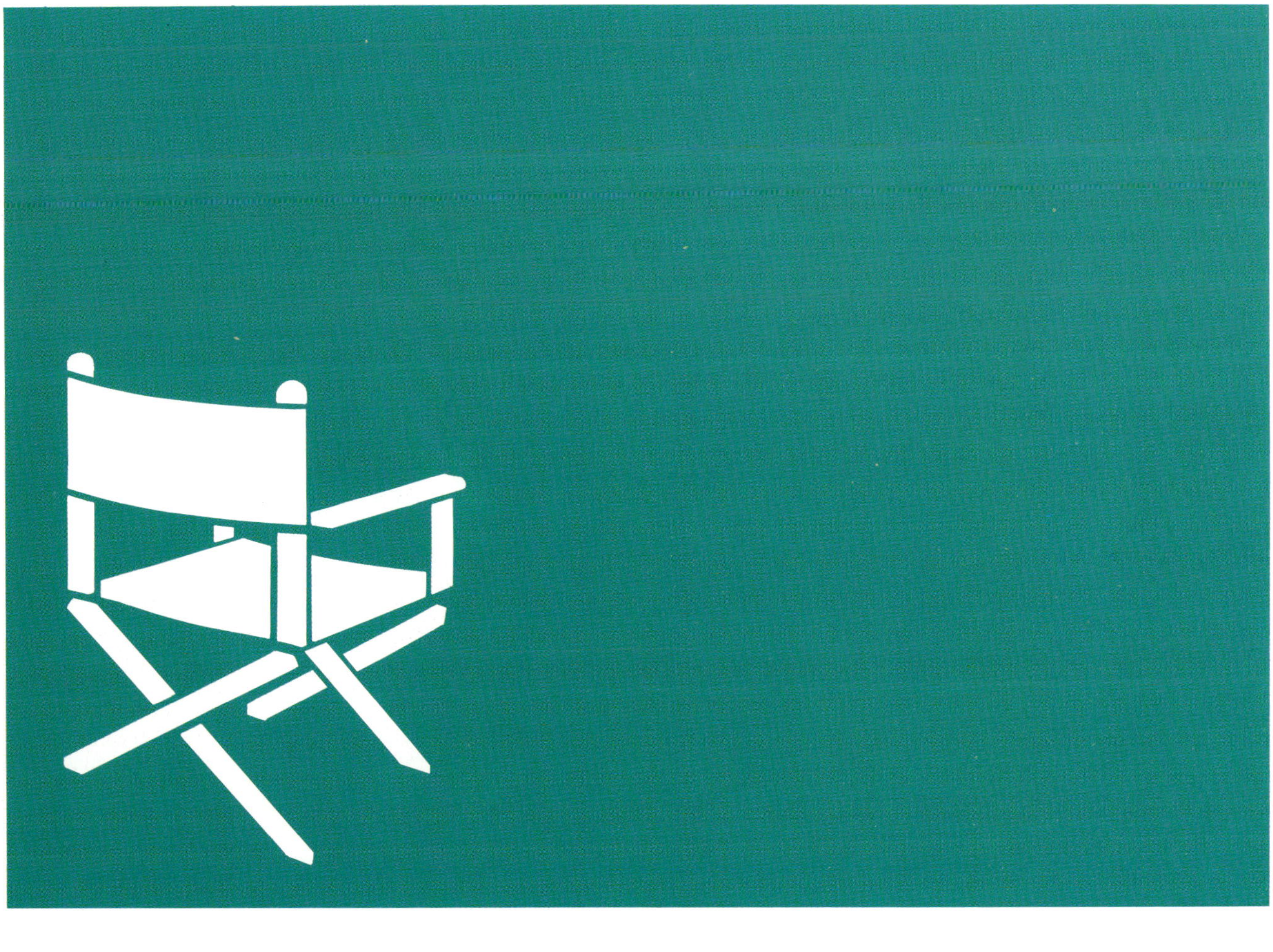

# Chapter 9
# Camera Angles

In Chapter 7 we considered camera angles and their relationship to continuity and the maintenance of correct screen directions. There is, however, much more to the choice of angles than these purely functional considerations.

By our selection of a certain angle we force viewers to see the subject from a particular point of view, and in its turn the point of view can often influence the way they regard the subject as they look at it. The viewpoint can add an extra dimension of meaning to the viewing experience — a psychological and dramatic one. But before we move on to consider what might be called the unusual angles, let's just start by thinking about the 'normal' one.

This is the one that newcomers to video making tend to adopt instinctively when they first acquire a camcorder; they hold it at eye level and shoot the subject broadside or head on. There isn't actually anything wrong about this; it's just that in many cases a more thoughtful and enterprising choice of angle would add extra visual (and often dramatic) interest to the shot.

*A head-on view of a subject (left) often gives a flat and uninteresting rendering. If, however, we move the camera slightly to one side for an oblique angle (right), the subject is immediately given a three-dimensional solidity because its depth is then revealed.*

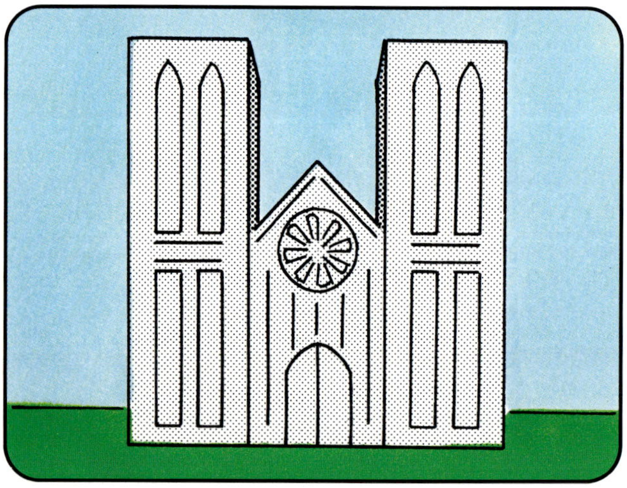

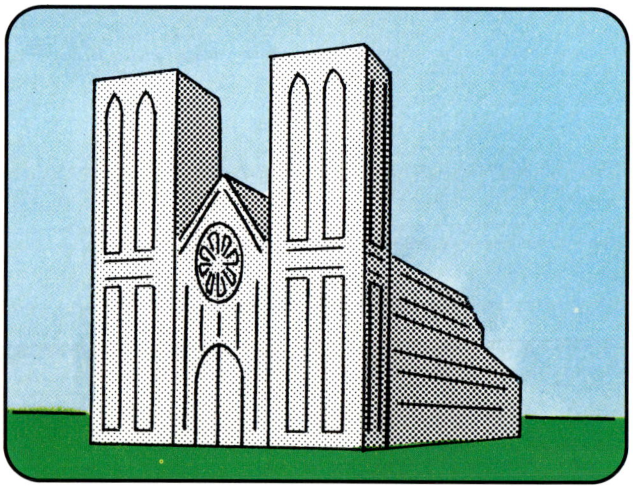

By definition such a normal angle must be ordinary; it's a neutral kind of shot that is in effect presenting the subject without the camera making any comment of its own.

## Oblique angles

As a general rule, you can say that the view of a subject from an oblique angle (that is, one obtained from a position slightly to the side) is likely to be more interesting than a shot that's taken head on. This is because the oblique angle allows us to see something of the side of the subject as well as the front and so gives an impression of its depth.

This is true for pretty well any kind of subject from steam locomotives and cathedrals to the human face. In the case of the latter, it also tends to provide the most flattering portrait. The angled view, with the subject's face half turned towards the camera, will help diminish the effect of such shortcomings as protruding ears, close-set eyes and a mouth that's a trifle on the wide side.

At the same time, while a profile shot would obviously emphasise an over-long nose, the angled view helps restore the offending feature to more elegant proportions.

There's also a sound technical reason for using an oblique angle in certain shots which would otherwise have a direct lateral movement. With some combinations of camera distance and speed of movement, subjects with strongly marked vertical features can produce an unpleasant 'juddering' effect when they cross the screen if shot with the camera broadside on. The use of an oblique angle will avoid this nasty visual sensation.

## High angles

A high angle shot means that the camera is placed in an elevated position so that it looks down on the subject and gives the viewer a feeling of dominance and superiority. For this reason it's often used in a dramatic context as a subjective or point-of-view shot of a character who is dominating or in some way gaining an advantage over another.

As the camera is directed down, the apparent height of the subject is diminished because of the foreshortening effect of perspective, and so it tends to look less impressive.

Such a shot would naturally be taken fairly close to the subject, but at greater camera distances the high angle has the effect of emphasising the location. This is because the aerial viewpoint reveals its extent.

The ultimate high angle is the overhead shot which is occasionally used to give a stylised, 'diagrammatic' look to the action. This is a specialised and fairly rare camera technique. It's particularly effective where crowds are involved; they can melt away to reveal a person or object lying on the ground, or be used to show a chase when a wave-like movement of jostled bodies charts the progress of the pursuer and pursued.

Overhead shots have often been used to good effect in stairwell sequences. The spiralling bannister rail not only creates an interesting visual pattern, but the viewpoint gives an air of eeriness and mystery to characters going up or down the stairs because the unnatural angle, by concealing their features, robs them temporarily of their identity.

Of course professional movie makers are able to use special equipment, such as camera cranes, to obtain their extreme high angles — facilities that are not normally available to amateurs. However, it's amazing what can often be done by making use of features which your chosen location may offer. Out in the country, convenient elevated camera positions may be found on top of hillocks or the easily accessible fork of some sturdy tree (though be careful here for the sake of limbs and apparatus!). In houses there are landings and staircases, or balconies in local halls. For moderate high angles, a step ladder will probably be all you need.

## Low angles

As might be expected, the low angle with the camera looking up produces precisely the opposite visual and psychological effects to those achieved by a high viewpoint. Whereas a character is made to appear diminished and less important by being shot from a high angle, a low angle will make him or her into a dominating figure which towers above the spectator. For this reason it's very often used to give the subjective viewpoint of a victim, suppliant or anyone under the thumb of a dominant character, and generally as a reverse angle to the elevated p-o-v shot already discussed.

Low angles can also be used to emphasise the height of structures which are tall anyway. I should mention, though, that the success of this device is dependent on the subject having some strongly marked vertical lines in it (such as a

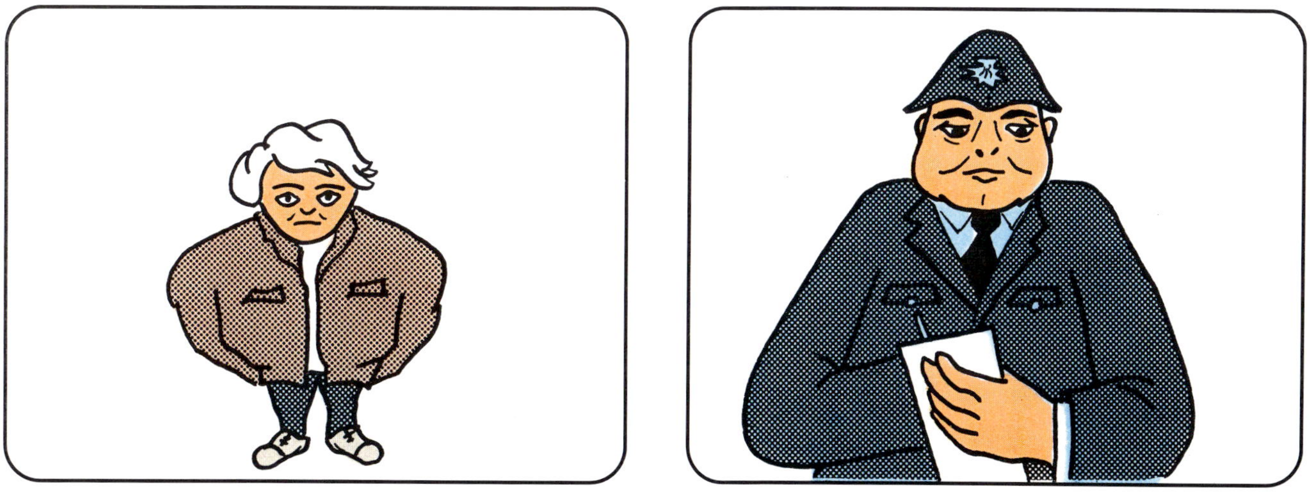

*A high-angle (left) diminishes the subject and makes it seem less significant, while an upward-looking low angle (right) causes the subject to loom large in the frame and dominate.*

skyscraper), because it's the sharp convergence of these through perspective that's largely responsible for the dramatic effect. If, for instance, one turns the camera upwards at a featureless cliff face, the result on screen is likely to be disappointing.

As I've said before (but it *is* an important point), tilting the camera skywards can cause exposure problems unless some manual control is applied to the aperture to compensate for the excess light influencing the meter, and we must make sure that direct sunlight never strikes the camera's sensitive image sensor as permanent damage can be done.

So far we've only considered low viewpoints in terms of the camera shooting at an upward angle, but this is just half the story. Low angles can be equally useful when the camera is aligned to shoot at an angle more or less parallel to the ground. The point is that when the eye-level of the viewer is reduced to something like ground level, the terrain itself virtually disappears and all objects, near and far, seem to be on a common horizon line. This has the effect of making for bold uncluttered compositions by causing the action (at least, in open country) to stand out against a plain sky background.

One use for the low-angle shot involves that traditional amateur movie subject — baby-on-the-lawn. Toddling infants are, by their nature, extremely short subjects, so if we attempt to shoot them from adult eye level it's necessary to tilt the camera down at a steep angle. Result: unattractive foreshortening of baby and rather uninteresting background of lawn.

Once you have established the garden location, a much better way to tackle the task of recording baby's first faltering steps is to get down to the toddler's own level. Shooting from this angle will eliminate most of the dull lawn and may even bring some more interesting features of the scene into frame. But most important, the picture will now be dominated by Baby — the reason for the whole operation.

### Composition

It seems appropriate to finish this chapter on camera angles with a few general comments about pictorial composition. In the realms of

*The Rule of Thirds in composition. When the scene in the viewfinder is divided into three sections, vertically and horizontally, the main features of interest should, as far as possible, be placed at the intersections of the imaginary dividing lines. In the case of the long shot (left) these are the yacht and the lighthouse lantern, and in the big close-up (right) the character's eyes.*

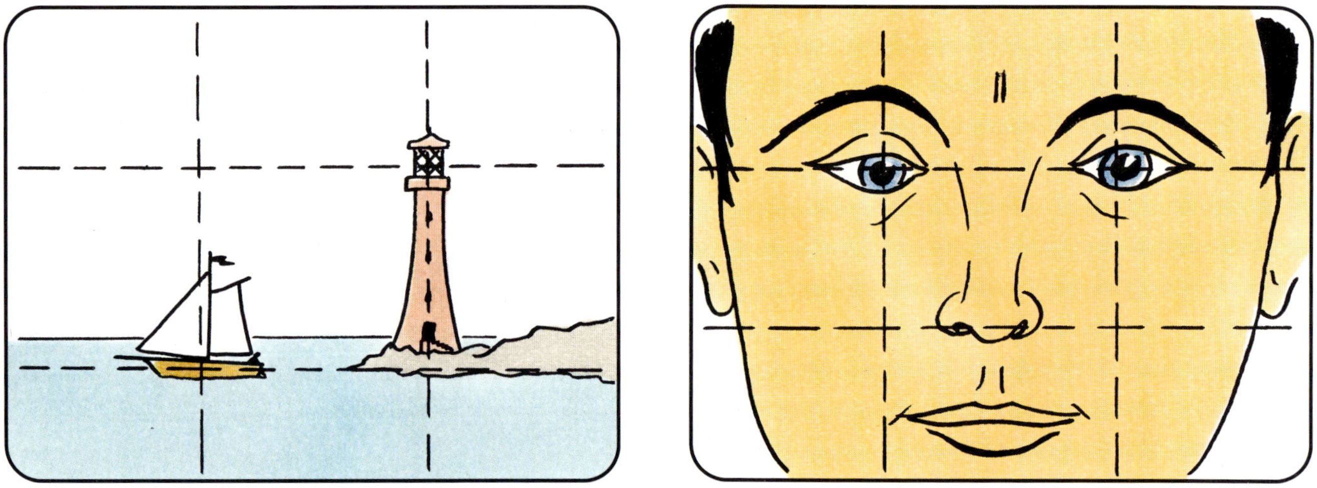

*When composing shots look out for background features which could cause unwanted laughs — such as smoke pouring out of the subject's hat! Adjust the camera position so that the source of the smoke, if it can't be excluded, becomes part of the composition.*

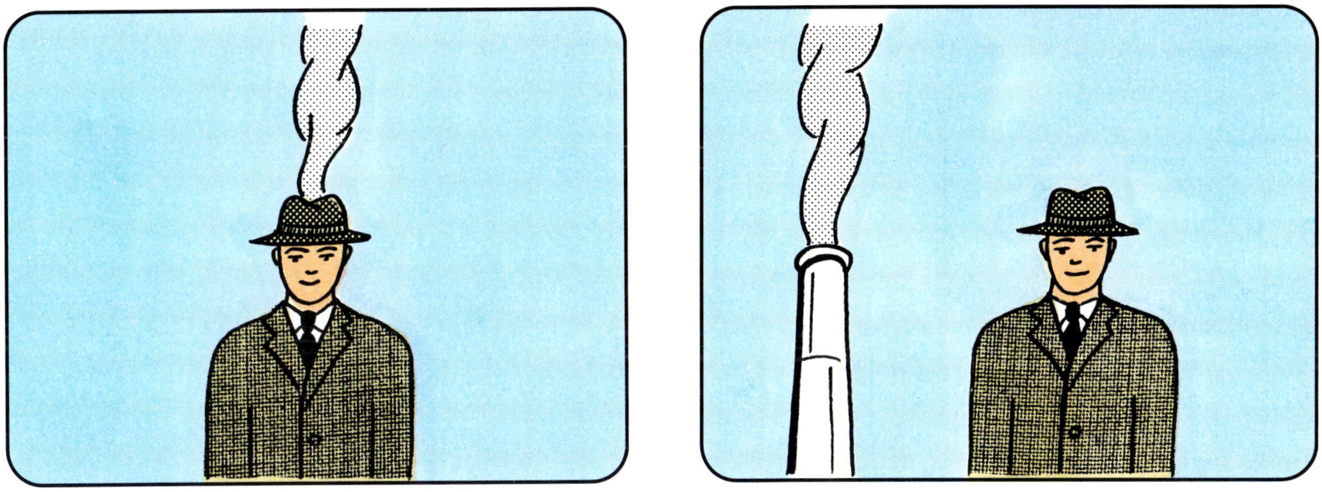

painting, drawing and photography, composition is something of an exact science. This is possible because these arts are dealing with *still* images whose form can be constructed and analysed in terms of fixed geometrical patterns.

However, in video making we are dealing with constantly moving images so the classic rules have to be relaxed and adapted to accommodate these. All the same, many of the basic principles of composition can still prove useful when there is not too much frantic activity going on within the frame.

Probably the most helpful of these is known as the Rule of Thirds. This teaches that we should mentally divide the picture area into three sections, horizontally and vertically; we should then place the main features of interest as close as possible to the four points where our imaginary dividing lines cross (see illustration). In a seascape shot, for instance, we might have a yacht in the lower left-hand side of the frame and a lighthouse on the right; here we would get the most balanced effect if we were to compose the picture so that the yacht was positioned at the lower left intersection point, with the lantern of the lighthouse (its main feature of interest) at the upper intersection point towards the top right of the frame.

Then again, in a big close-up of a face the features should be placed within the frame so that the eyes occupy approximately the positions of the two upper intersection points. This will obviously mean that the top of the head is cut off, but this doesn't matter because the other main feature — the mouth — will be included.

While on the subject of close-ups, we should always remember to allow 'looking space' when showing a character watching some action that's taking place off screen. This means that instead of placing the head of the character right in the middle of the frame, we set it slightly to one side. For example, if the character is looking to the right, the picture is composed with the head slightly over to the left of frame so that there is more space (the 'looking space' referred to) in front of it than behind. This gives the shot a much more pleasing appearance than if the character were staring up against the blank edge of the screen.

On the other hand, if a close-up face is looking forward straight at the camera (and so directly at the viewer), it's usually best to place this centre of frame, unless there's some other compositional element that has to be included in the picture.

When shooting a group of four or five people,

9

A low-angle shot of a tall building, especially when taken on a wide-angle lens setting, emphasises the impression of height. Note how the converging vertical lines in the subject enhance the effect.

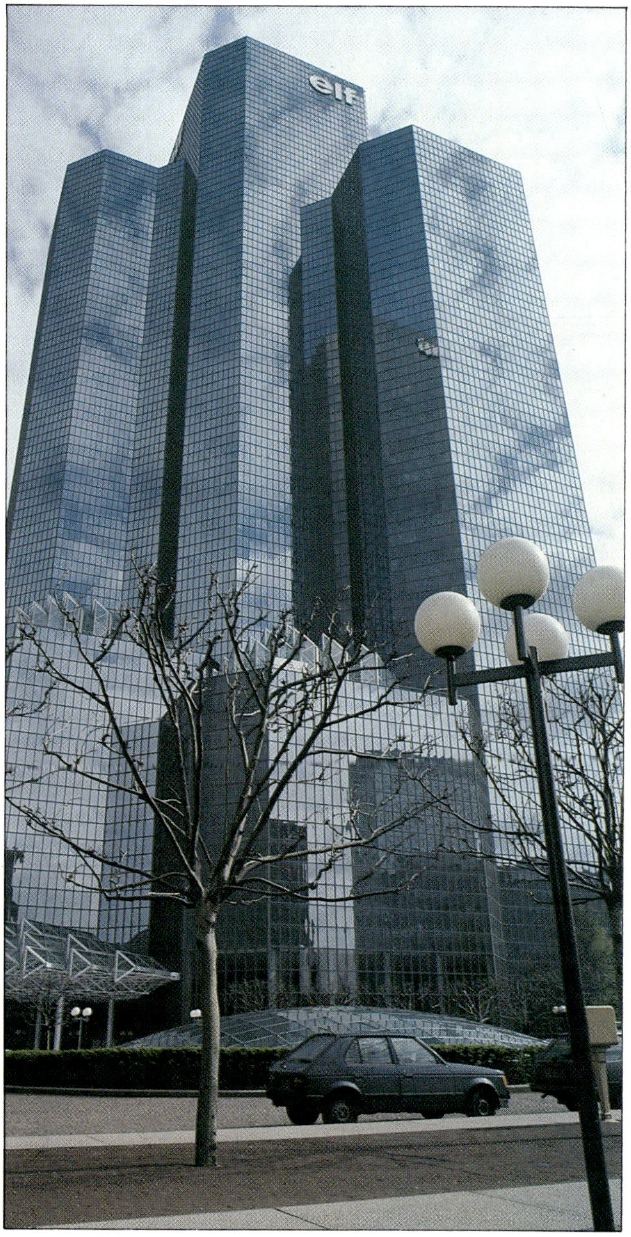


remember what I said earlier about an oblique angle usually being more interesting than a 'broadside'. If they're just lined up in front of the camera they'll look stiff and unnatural; stagger them slightly and take the shot at an angle of about 45 degrees.

Another point to watch for when composing a shot is some distracting feature in the background. If you're not careful, you may get some unintentionally comic effect such as smoke pouring out of the top of somebody's head because there's a factory chimney directly behind them in the distance. In such a case adjust the camera/subject position so that the chimney, if it can't be totally excluded, is clearly to one side.

Although it's necessary before a Sunday dinner, we certainly mustn't 'carve the joint' when composing our video shots. By this I mean that when showing less than the full human figure in a medium or close shot, we shouldn't cut it off exactly at the ankles, knees, waist or neck. Make sure instead that the lower edge of the frame slices across at some intermediate point, say, just above the knees in mid shot or below the shoulders in a close-up.

To sum up on composition, the overriding consideration should be that the way we arrange the picture elements within the frame should always lead the viewer's eye naturally to the feature that we wish it to see first. For example, the converging lines of a straight road running from foreground to horizon would guide the eye immediately to a figure walking along it in the far distance. And we must always remember that, however we compose the picture, it is *movement* that will gain attention before anything else.

# Chapter 10
# **Lighting**

There's much more to lighting a scene than merely mustering enough lumens to record a recognisable image on tape. Without, I hope, sounding too pretentious, we can say that light is the 'paint' with which video makers create their pictures, and like any other artist's material it needs to be used with understanding to get the best results.

### Colour temperature

But before getting down to the basics of how using light intelligently can help us to make better videos, it's important to remember that it comes in two separate varieties — natural and artificial. This is turn brings us to the question of *colour temperature* and the effect it has upon the way colours are rendered.

It's an unfortunate fact or life that cameras — electronic of photographic — don't possess the tolerance of the human eye when it comes to registering colours lit by different types of light. We might look at a person's face in the sun then again indoors by electric light and there wouldn't appear to be very much difference in the flesh tones. However, if we were to take video shots of that same person and didn't make appropriate allowance for the particular qualities of the sun and artificial light, we would notice a marked difference in the results when these were played back. Shooting by tungsten lighting with the camera adjusted for daylight would cause the flesh tones to assume a very unnatural orange-yellow tinge.

This is because the sun and electric light have different colour temperatures, which are expressed in degrees Kelvin. Sunlight (at noon) is reckoned as 5,500K, while photographic lamps (when new) are rated at 3,400K. We don't need to worry about these actual figures; the important thing to remember is that the *higher* the colour temperature the *bluer* the light, and the *lower* the *redder*.

So you will see from the figures quoted above that sunlight has more blue in it and artificial light more red, which is why recorded colours look different by the two types of illumination and why special measures have to be taken if hues are to appear 'natural' (perhaps a relative term in colour photography) in each.

To cope with this problem, video cameras have a daylight and tungsten setting switch and a *white balance* control which ensures that colour renderings will be accurate, whether the scene is being illuminated by natural or artificial light. It is of course essential to ensure that these controls have been set for the prevailing lighting before a shot is taken.

*There are two possible solutions to the colour temperature problem when shooting in a mixture of artificial and natural light. One technique (left) is to shoot for artificial light, with the camera's controls set for tungsten illumination, and place appropriate orange filter material over the window to modify the colour temperature of the daylight. Alternatively, you can shoot for natural light and match the lamp to the daylight by placing a blue filter in front. In this case, the camera will have to be set for daylight shooting.*

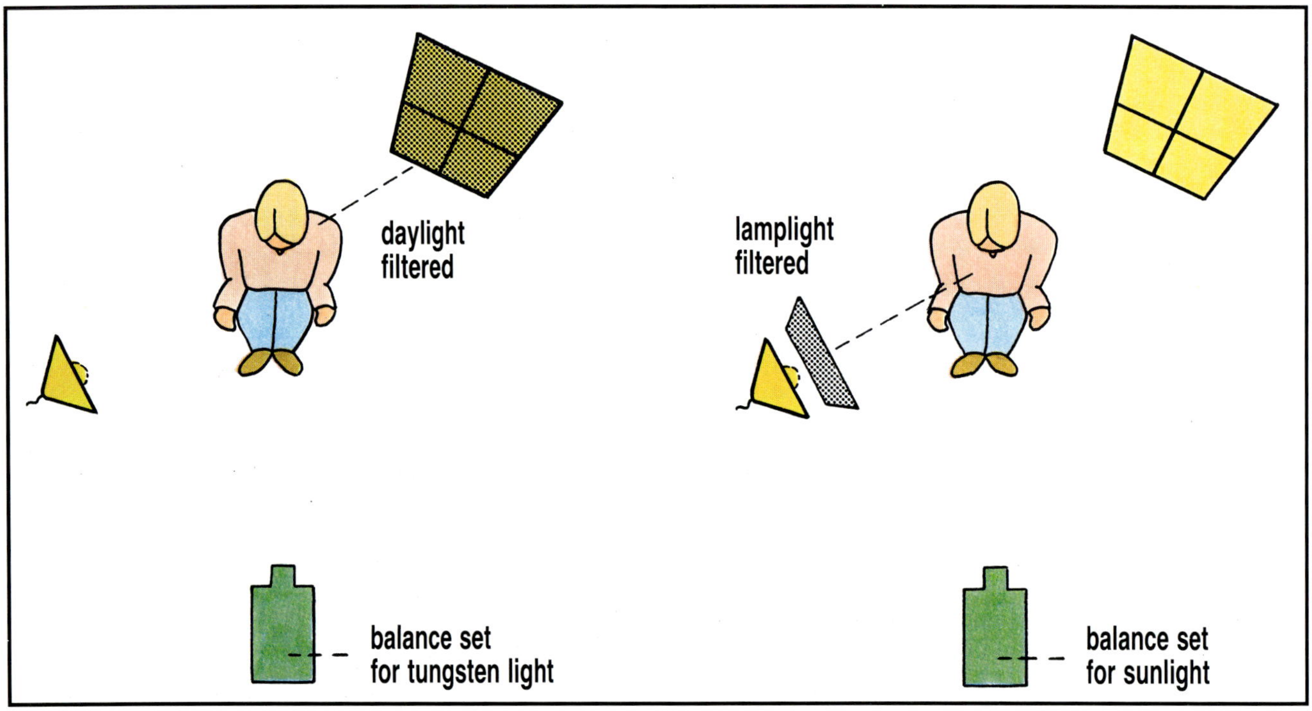

*Shooting in bright sunshine can have its problems. If the sun is right behind the camera (set-up A) faces look flat, with a shadow under the nose like a toothbrush moustache; also, subjects tend to screw up their eyes against the glare. Moving so that the sun is slightly to one side (set-up B) improves things: the side shadow 'models' the face to give it solidity and the subject looks more relaxed. However, the marked contrast between light and shade is rather unattractive in a close-up, so ideally a reflector should be used (set-up C) to lighten the shadow for a more pleasing effect.*

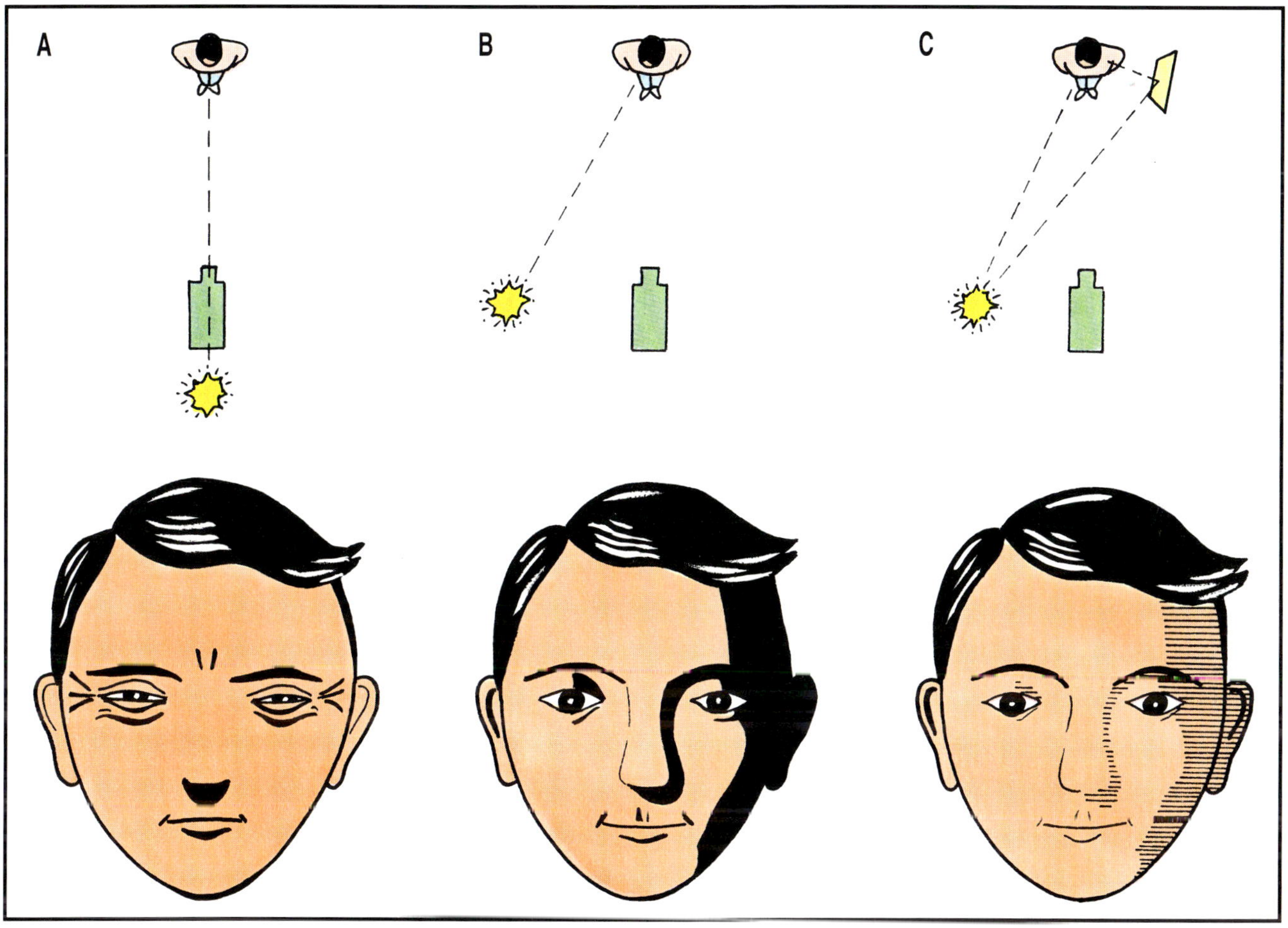


As I mentioned above, a higher colour temperature means the light is bluer than with a lower one. This sometimes causes confusion because it seems more natural to speak of a picture having a 'warm' look if it has a rosy glow about it, while blue is often thought of as being a cold colour. But the way to remember it is to think of a piece of metal that's white hot (and so emitting more blue light) as being hotter than one that's only red hot.

### The mixed light problem

This matter of using the correct sunlight/tungsten and white balance setting is all pretty straightforward if we're shooting in *either* daylight or lamplight, but what happens if there's a mixture of both?

Perhaps we're shooting a character in a room and because of the low light level need to use artificial illumination. We set up our lights and adjust the camera for tungsten — but there's a window in shot showing a sunlit garden scene outside. If we go ahead and shoot as things are, this window scene will have an unpleasant blue cast that we certainly don't want. So what can we do?

There are two possible solutions: we can either filter the lamps with blue gelatins to match the colour temperature of the window scene and shoot for daylight; or cover the window with

10

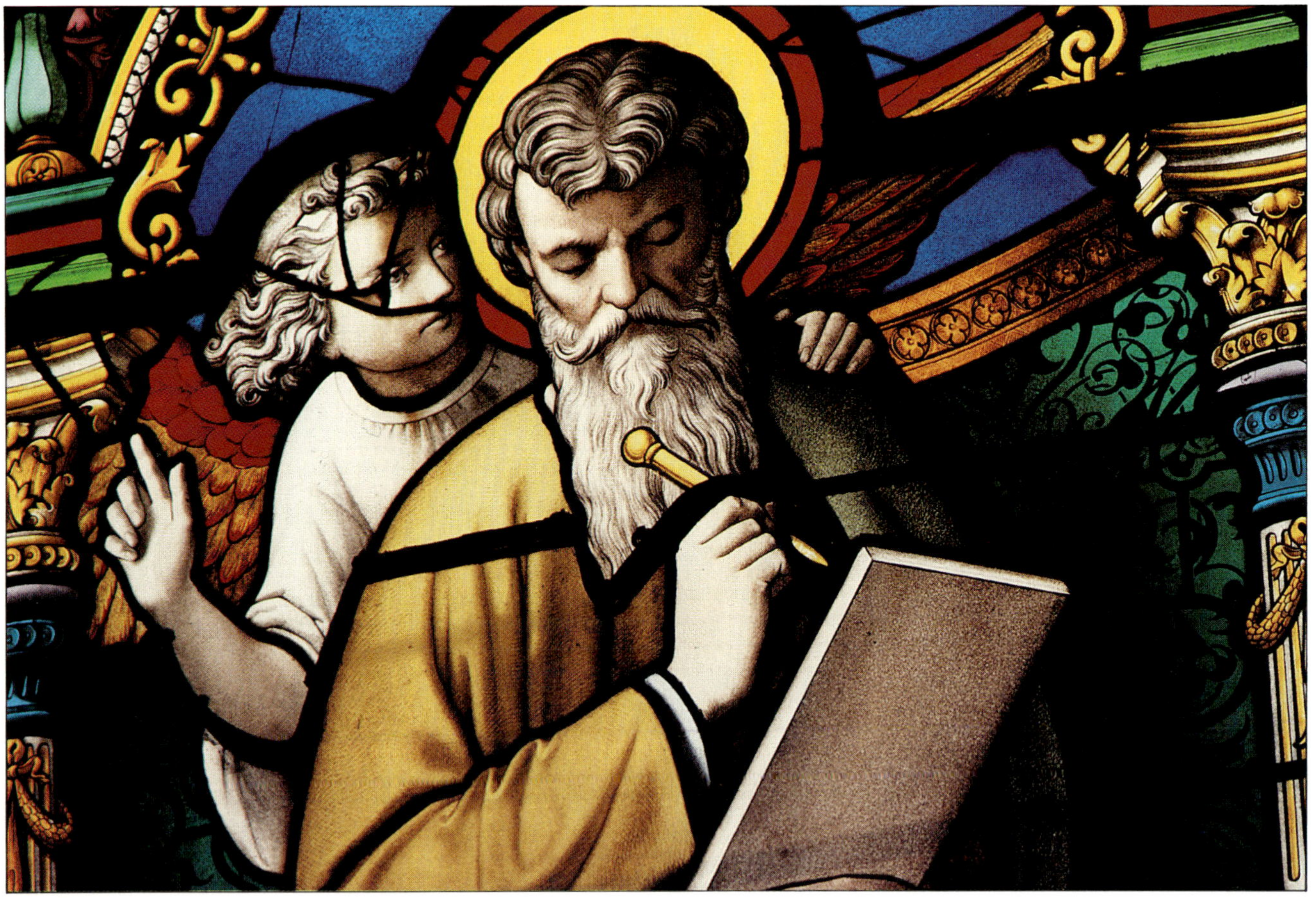

*A translucent subject, such as a stained glass window, which is shot by transmitted light — that is, light which is coming through it — always records with rich, bright colours. Balloons, plastic umbrellas and sails of windsurfing boards also look good recorded this way.*

orange filter material to match the lamps and shoot for artificial light. Both these methods have their snags. Filters on the lamps will reduce their efficiency, while covering a large window could prove inconvenient and expensive, so which solution is best will depend rather on your individual circumstances.

Having considered the technicalities of the two types of light, let's move on now to look at how they may be used for best artistic effect.

### Shooting in daylight

This is what most of us do most of the time; all those subjects so beloved of amateur movie makers — baby-on-the-lawn, picnics, sports events, holidays — are recorded by the light of the sun in some form or other.

Contrary to what is often supposed by beginners, a bright sun in a cloudless sky does *not* provide the best light for photography. Such conditions produce harsh shadows, and a degree of contrast between these and brightly lit areas of the scene which are hard for the video camera to handle successfully.

In this situation, what you should never do is shoot with the sun directly behind the camera; this will produce a flat, uninteresting effect and if the subject is human will have those involved screwing up their eyes against the glare. Move so that the sun is to one side or the other at an angle of about forty-five degrees to the camera's line of fire. This will add solidity to the subject because some side shadow will now be visible (what photographers call *modelling*) and will save your characters from having to squint directly into the sun.

But although shadows provide this desirable modelling, they can look very unflattering in a close-up if too hard. The answer here is to use a reflector on the shadow side just out of camera range. It's surprising what an improvement can be achieved by just holding up a sheet of newspaper — and you probably won't have anything more elaborate with you while movie making on holiday.

However, when one's able to take time over a set-up, more sophisticated reflectors with matt-white or shiny surfaces can produce subtler effects. They are particularly useful for backlit close-ups in bright sunshine. If you're wondering why anyone would want to take such a shot, it's because the technique produces an attractive rim of highlight around the subject's hair, while giving a helpful degree of separation between subject and background — and, of course, there are no harsh shadows on the face which is being lit by reflected sunlight.

There's another special case when backlighting

*A basic studio lighting set-up. The main source of illumination on the subject is the key light, A, which is angled at approximately 45 degrees to the camera lens axis. B is the fill light to soften the shadows on the other side of the subject; this is also placed at about 45 degrees to the camera, but further away than the key light (a more pleasing effect is sometimes obtained if a diffusion filter is put in front of the fill light). The spotlight, C, is used to backlight the subject from above; this helps to 'lift' the subject away from the background by providing a separating rim of highlight around the hair. D is a floodlight on the background, its main function being to 'kill' any ugly multiple shadows produced by A and B.*

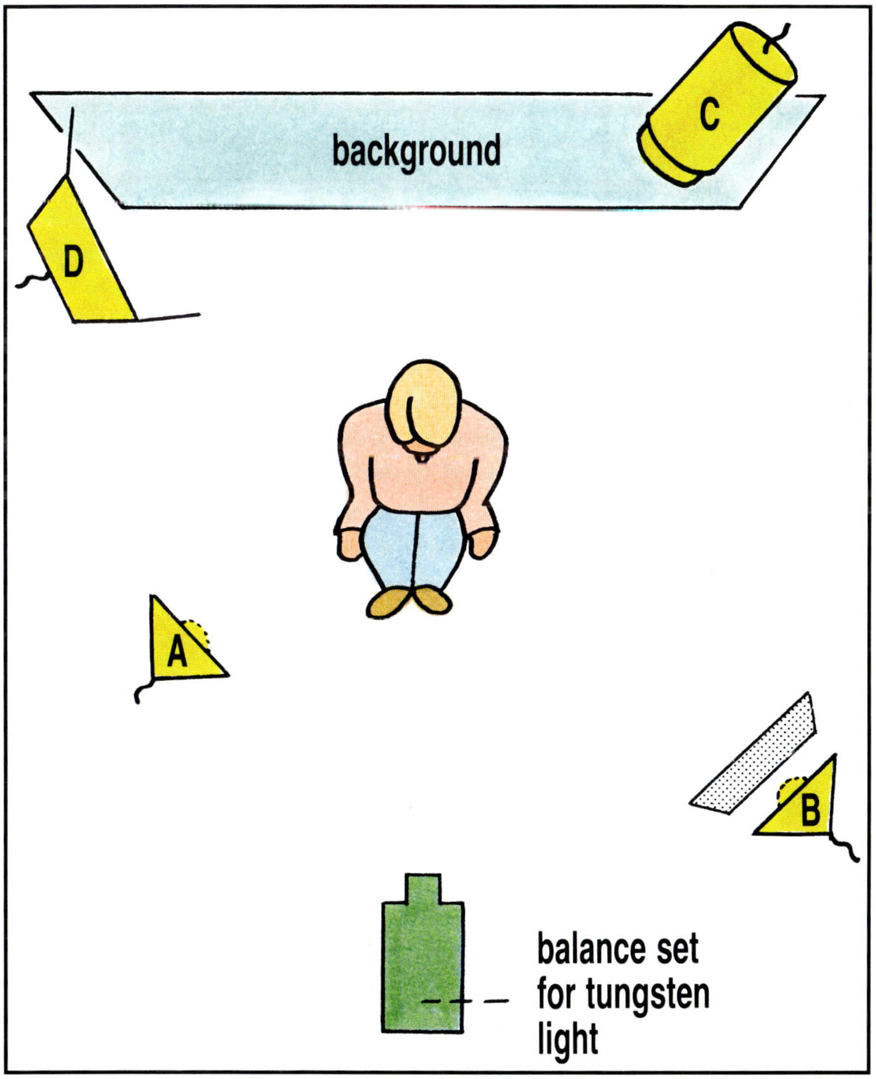

comes into its own — the recording of translucent subjects. An obvious example is stained glass windows from inside a church, but the technique can also work in the open with things like the sails of windsurfing boards, coloured umbrellas and toy balloons. The colours really appear to glow because *transmitted* light is always more intense than light that's being merely reflected.

The most useful natural lighting for movie makers is, fortunately, the sort we tend to get a lot of — the condition known as *cloudy-bright.* Here the sun is thinly veiled by cloud, so its light is diffused and there are no harsh shadows. Under these conditions, we are free to shoot from just about any angle we please, and can happily leave the camera's automatic meter to take care of the exposure without having to worry about whether we should be pressing the backlight button.

## Shooting by artificial light

Artificial lighting for movies might be said to fall into two categories: *functional* for newsreel-type reportage, and *studio* when we have everything under our control and can light the subject for the most artistic result.

The functional type usually means one videolight mounted on a camera bracket, and it will allow you to get usable results under rough-and-ready shooting conditions, but without much finesse. The main trouble is that the lighting tends to be harsh and contrasty, and it produces dark shadows which bob about behind the subject if it's close to a light background. If there is no background, the area to the rear disappears into a murky void. Another snag is that the single videolight, like the sun in a front-lit shot, can dazzle human subjects and put them off their stroke.

These problems can be lessened to a certain extent by taking the light off the camera mounting and placing it on a stand (or having it held by an assistant) as high as possible, to one side of the camera. So doing will make the shadows less obtrusive — the higher placing of the lamp will cause them to be cast lower down the background — and the angled illumination reduces the 'flat' look. You will also avoid the changes in light intensity as the lamp-to-subject distance alters with different camera positions.

Another method is to use *bounce lighting.* This can be done where the action is being shot in a room with a white ceiling or walls, and where the subject area is not extensive — say, a close medium shot of a character sitting in a chair. The single videolight is simply directed at the ceiling (or wall) which then acts as a large reflector and illuminates the subject with a diffused, even light. This technique can work well when the camera is switched to its low-light setting.

As you gain experience with artificial light shooting, you will possibly wish for more polished results by using a number of photographic lamps in the studio manner. Whole books have been written about such lighting techniques and they're certainly essential reading if you intend to do serious work of this kind. But in essence they involve the use of a *keylight* (for main illumination of the subject), a *fill light* (to lighten the shadows caused by the keylight) and backlighting (usually a spotlight to 'lift' the subject away from the background by adding a rim of highlight to the hair). There may also be a need to floodlight the background to 'kill' unwanted multiple shadows caused by the other lamps.

But a word of warning: never be tempted to use photographic bulbs for shooting by putting them in ordinary table lamps or ceiling fittings because they operate at very high temperatures and can be dangerous in normal sockets and under shades.

# Chapter 11
# **Editing**

The very fact that movies are made up of individual shots implies that these separate segments of action need to be linked together in a logical way — the procedure known as *editing*. This is regarded by many as the most creative aspect of the whole movie making process.

So editing is much more than just a mechanical business of linking one shot to another as though making a pictorial paperchain. It involves not only assembling, but also selecting images and deciding how long they shall remain on the screen. It's at the editing stage that the pace and shape of a movie are determined — indeed, its success or failure largely depends on the skill with which this crucial operation is carried out.

There are two distinct approaches to video editing: *editing in camera* and *post-production editing*. The first, as the term implies, is actually carried out at the time of shooting, while the other is done after the event once all the required material has been recorded.

# 11

## Editing in camera

Some purists say that editing in camera is really a contradiction in terms because editing is essentially a process of selecting and arranging existing material. While this is, of course, true in a strict sense, I think it's fair to say that video makers *can* be selective and pre-arrange their shots at the time of shooting. So this method is, in many cases, a valid way of working.

It's often suggested to be a suitable method for beginners because it requires no special equipment (apart from a camcorder that gives shot-to-shot transitions without picture disturbance) and is therefore thought to be easy. However, although it's undoubtedly technically simple, to be effective, it does require a degree of skill that's unlikely to be possessed by the inexperienced worker.

The whole secret of success with in-camera editing is planning and thinking before you shoot. For a start, as there will be no subsequent rearrangement of the material, all the shots have to be taken in the order in which they will appear in the finished production. This, as a moment's thought will show, can present certain practical difficulties. If, for example, we had a sequence of three shots, the first taken from the bottom of a cliff, the second from the top and the third from the bottom again, it would obviously be necessary to climb to the top of the cliff and down again between shots one and three — which could be tiring and time-consuming to say the least!

The in-camera editor also needs to be very clear in his mind as to when to start and end a shot as there can be no 'topping and tailing' later on (as is possible with post-production editing). For this reason it may be necessary to rehearse a piece of action in advance so that he can have a clear idea of the best in and out points before pressing the camera release.

Speaking of the start and end of shots, there's one very important point we have to bear in mind here. To achieve clean cuts without picture disturbance at shot changes, camcorders and recorders use a system known as *backspace editing*. This works on the principle of rolling back the tape a short distance when the camcorder is put in the record pause mode. At the start of the next shot, this section of tape is run forward again to give the mechanism a chance to lock on to the previous shot's control pulses. Only when this has happened does recording actually begin, which means in practice that we lose the first 1½ seconds of the new shot unless we press the release a little early to allow for the run-up time.

As I've emphasised earlier in this book, the movie maker *must* think in terms of sequences rather than individual shots, so the in-camera editor needs to have the overall shape of a sequence clearly in mind before shooting anything at all; only in this way will it be possible to achieve the essential feeling of logical progression from shot to shot that's needed to bind the whole thing together.

From what I've said, I expect you'll see that successful in-camera editing is something that only comes with experience, so, if you're a beginner, don't be too disheartened if your first attempts aren't as good as you had hoped. This is an area where practice makes your results, if not exactly perfect, at least a great deal better.

## Post-production editing

Because of the nature of the medium, video editing is essentially a re-recording or *dubbing* process. This means that at least two VCRs and a TV monitor are necessary to do it. One machine plays back the original material, the second records the edited version and the monitor enables us to see what we're doing.

As the final edited version is a copy or *second generation*, as it's called, it loses a little of its original quality — at least, when the process is carried out on domestic equipment. Such loss can be minimised to a certain extent by using the best equipment at our disposal, the highest quality tape and the standard (as opposed to the long play) speed when making the original recording.

There are also units called *enhancers* which can often help, if connected between the playback and copying VCRs.

The basic technique of editing is to feed the signal from the recorder that's playing back the original material (call it VCR A) into the machine making the editing copy (call it VCR B). Naturally, only the required material is transferred to VCR B and in this way the edited version of the programme is built up section by section, a process called *assemble editing*. In practice, a camcorder, if it has the necessary playback facilities, could be used instead of VCR A.

The TV monitor is connected to the output of VCR B. This arrangement means that when VCR B is in the play mode, its picture will show on the screen, but when it's on stop or record, the picture from VCR A (the original material) will be displayed.

The connections between the two VCRs should be from and to the respective video and sound out and in sockets, and *not* the RF connections which, although they would work after a fashion, give inferior quality — and we need all the reserve of quality we can get.

It's worth pointing out that the two VCRs don't necessarily have to be of the same format. The practical advantage of this is that if you have access to a U-matic recorder, you can make an edited master copy on this larger format, which

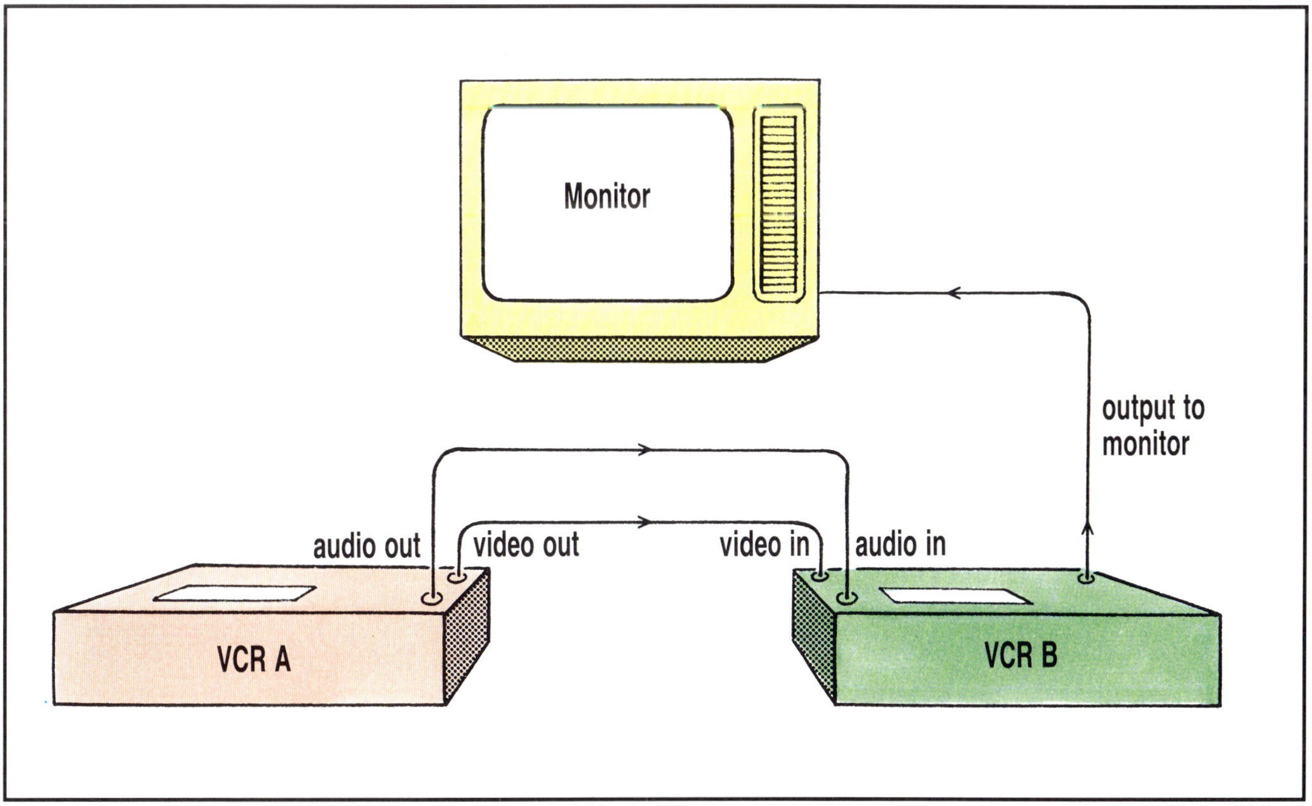


*The basic set-up needed for video editing. VCR A carries the original material which is fed into the copying recorder, VCR B, via the appropriate audio and video connections (the RF connections should not be used because quality would be lost). The monitor, connected to the output of VCR B, will show the programme coming from A when B is on stop or in the record mode.*

would be the thing to do if you wished to have a number of further copies made of the edited version.

For the most precise editing, both VCRs should have slow-motion, still picture and frame advance, as these facilities will help in the location of cutting points — which can, of course, only be identified while the picture is on the monitor. The essential requirement for recording VCR B is the backspace edit feature needed if cuts between scenes are to be free of picture disturbance. It's also desirable that B should be capable of doing *insert edits* for greater flexibility.

### Insert editing

The insert edit facility allows new material (with domestic equipment normally picture only) to be dropped into a pre-recorded tape without picture disturbance at the in and out points. It's particularly useful for inserting cut-away shots into long continuous takes to add visual variety. For example, we might have an extended take of a speaker addressing a meeting, and by insert editing we could include shots of listeners while the speech continued uninterrupted on the soundtrack.

The amateur video editor is up against the problem that he can't be quite sure in advance precisely which frames of the outgoing and incoming shots will be 'spliced' together. It's not just a matter of finding the required frame,

*How to allow for the backspacing principle when editing. The tape on the recording VCR B backspaces from the end of the outgoing shot, point v, to x, then goes into record with the incoming shot at w. To compensate for this, the actual tape start point of the incoming shot, z, on feed VCR A must be set* before *the required edit in point, y, by the distance yz, while the tape start point of the outgoing shot on VCR B must be set* after *the required edit out point, v, by the distance vw. In this way the loss of the first 1½sec (approx) of the incoming shot and the last few frames of the outgoing shot is avoided.*

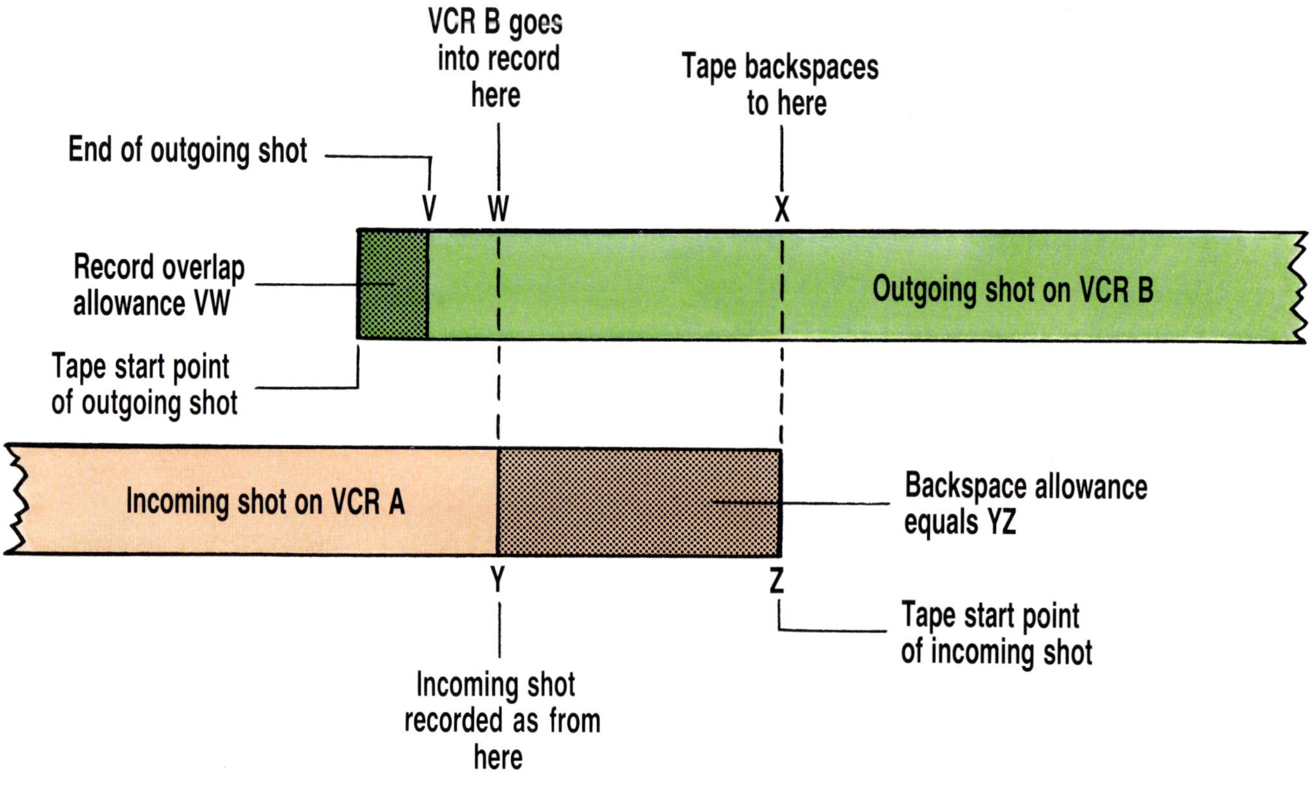

holding it on pause and then starting the playback and record machines together. Unfortunately life's not that simple.

Remember the backspace editing principle I mentioned earlier under *Editing in Camera?* Once again, this roll-back time (usually about 1½sec) must be allowed for when editing, otherwise the first moment of the incoming shot will be lost.

There is another subtlety connected with the backspace editing mechanism which must be taken into account; this affects the end of the outgoing shot (already recorded on the tape of VCR B). Things are so arranged that the VCR goes into record *just before* it reaches the end of the rolled back section of tape, which means that the last fraction of the outgoing shot will be recorded over and so lost. In practice, this is likely to be only a few frames but could make all the difference if a really precise cut is being attempted. It was mainly for this reason that I said in Chapter 8 that cuts on action can be tricky for the user of domestic video equipment.

These problems of what we might call backspace imprecision can be overcome to a large extent if we experiment with our equipment before attempting any really serious editing. If we know in advance how much of an incoming and outgoing shot is lost through backspacing, we can make an appropriate allowance when lining up the tapes before making an edit. The losses can be determined by doing trial recordings and then comparing them with the original tape, using the frame advance facility to count frames to see how many are missing. When doing this it helps to use a shot with clearly defined action at the beginning and end.

If we find, say, that the run-up time on VCR B causes 37 frames to be lost from the incoming shot, we can then set the tape on VCR A 37 frames *before* the required edit in point so that it will be 'there' at the moment VCR B switches to record. Likewise, if we find the record overlap is causing six frames to be lost from the end of the outgoing shot, we can set the tape on VCR B six frames *after* the required edit out point to compensate for this when editing on the *next* shot.

To make the edit recording both VCRs are, of course, started together — you can help co-ordinate this action by counting down: 'Three, two, one, *start*'.

It's important, at the end of a transferred shot, to allow both playback and record machines to run on for a few seconds after the required out point has been reached. This means that there will be 'spare' frames in hand over which the next backspace edit can be made without the risk of any unwanted gaps between shots. Such extra frames are, of course, erased and recorded over by the next incoming shot, so it doesn't matter within reason how much additional coverage you allow.

## Finding the shots

Perhaps it hardly needs saying that before we can attempt to edit a collection of shots together we've got to be able to find the ones we want to use. This means that we must have some way of readily identifying their position on the cassette, a process usually referred to as *logging*. This simply means that when we review all the material at our disposal to decide what's going to be included, we make a list of the required shots and note the reading on the tape counter where each one starts.

It naturally makes life easier at the editing stage if shots have been taken in something like the order in which they will ultimately appear — although, as we've seen, this is obviously not always possible — because it will save a lot of fast winding backwards and forwards to find the desired section of tape. For similar convenience, it's preferable to have all the shots for one movie on the same cassette.

Having looked at the equipment and techniques involved in editing video tape, we're now ready to consider the far more interesting creative aspects

# 11

of the subject. It needs to be said at the outset that editing is not a process that can be considered in isolation. It's no good shooting haphazardly when we're on location in the hope that everything will magically come right when we start sorting through the material on our monitor screen. Some thought — correction — a great deal of thought should be given to how the shots will eventually be put together *before* the camcorder release is pressed. Obviously an editor can't be expected to make bricks without straw, and if the shots for building coherent sequences just aren't there to start with, there's not much chance of a well structured movie emerging.

But having said that, there's usually something you can do to make even the most untidy material more presentable, and it's this aspect of editing that we'll consider first.

### Unscripted material

A high proportion of movie makers first acquire equipment to record an important family event, particularly a rather special holiday. As beginners, they're unlikely to have done even the most rudimentary pre-planning, so the material they return with will be somewhat disorganised, to put it kindly. How can order be brought out of this pictorial chaos?

The first move is to check the total footage for its technical quality. While carrying out this check, it's a good idea to make verbal notes on a sound tape recorder (if available). This is easier than trying to scribble as the tape is running; you can say something like: 'Cut shot of harbour before starting to pan' without taking your eyes off the screen. Also, note the number indicated on the tape counter at any point where some such action needs to be taken, as this will save time later on when you need to locate it to make the edit.

When you've completed this quality-checking operation, any shots that are technically below par (badly focused, incorrect colour balance, etc) should be clearly listed for non-inclusion when the final assembly is made. I don't advise erasing them at this stage in case you accidentally overrun and remove the wanted beginning of the next shot.

Even when the raw material is run in the order in which the individual shots were taken, it will almost certainly seem a shapeless mess with little feeling of continuity. The editor's job is to try and absorb the overall content and see how a properly defined structure can be imposed on what, at first sight, may seem a visual and aural rag-bag. This is not quite as difficult as it may appear; the trick is to think positively and have a logical approach.

By way of illustration, let's take a typical record of a family holiday centred, say, on a Swiss lake resort. Let's assume we came back with the following material:

*General scenes of departure from London Airport, shots taken from plane during the flight, and shots of arrival at the resort hotel — total duration 8 minutes.*

*Family strolling on the deck of a steamer during a trip across the lake, shots of passing views and some close-ups of family members — total duration 10 minutes.*

*Family exploring main centre of resort, looking at shops, etc — total duration 10 minutes.*

*Family on excursion to Swiss mountains by train and cable car — total duration 30 minutes.*

*Various shots of a steamer leaving quay and travelling across the lake — total duration 4 minutes.*

The first thing that's apparent from the footage as it stands is that it has no feeling of building to a climax; the odd shots of the lake steamer after the much more 'meaty' mountain excursion episode is weak and ineffectual. On the other hand, the beginning's not so bad — obviously the material relating to the outward journey would come first, so we leave that as it is for the moment.

The two short sections devoted to the lake and its steamer are not very exciting in isolation but could be made to amount to something if combined, so we make a note to do this (I'll explain how later).

It would seem more logical for the family to be seen about the resort before taking a trip on the lake, so we'll move this sequence to a position between the journey material and the to-be-combined steamer shots.

The excursion, in terms of running time and general interest, is the main feature of the holiday so should come at the end to form a fitting climax to the movie.

Our proposed assembly, which should be noted down, now reads like this:

*Travel sequence — 8 minutes*
*Family exploring resort — 10 minutes*
*Combined lake steamer material — 14 minutes*
*Excursion — 30 minutes.*

You don't have to be an expert to see that this is a far better structure; the flow of action has more logic and a sense of moving to a climax — the trip up the mountains.

The next move is to look critically at the individual sequences to assess whether their interest is proportionate to their length. Could one or more be shortened without losing anything of importance? Just because we have recorded eight minutes of travelling scenes, it doesn't mean that it's a good idea to use them all. Maybe the balance of our movie would be improved if they were reduced to four or five minutes.

### Shot continuity

Having made our sequences lead logically from one to another, we now have to perform a similar and rather more tricky task with the individual shots themselves. As we saw in Chapter 8, the arch enemy of movie continuity is the jump-cut — caused when the camera is stopped and then restarted from the same viewpoint. A similarly jarring effect is produced when the same subject is shown engaged in totally unrelated action in two consecutive shots.

If one shot in our holiday movie shows the family, backs to camera, looking in a shop window and in the very next they are seen walking along the street, that would be a kind of jump-cut even though the camera angle and possibly the image size of the subject had been changed. As with the 'classic' jump-cut already discussed, the way to cure it is to put something between the two offending shots — a cut-away or a cut-in (again, see Chapter 8).

As our family has been established in a street location, more or less any activity going on around them would make a suitable cut-away. But as they were shown specifically looking at a shop window, a cut-in to the goods on display would, in this instance, be more appropriate. It's to meet just such an editing emergency as this that we should always remember, when on location, to take plenty of shots for possible later use as cut-aways or cut-ins.

As I've already mentioned, psychology has an important part to play in continuity by establishing a logical connection between shots. In a travel sequence it wouldn't be good technique to cut straight from a shot of a stationary car in a street to a travelling view taken from inside a moving one. However, if we insert a brief close-up of a traffic light changing from red to green, we have a much more acceptable progression.

Harking back to the opening journey sequence of our holiday movie, it may seem difficult at first to link the shots of ground activity at the airport with those taken from the plane itself. However, as it's an introductory sequence it would be quite in order to bridge any awkward gaps with opening titles (either explanatory or credit).

If you feel that would be dodging the problem, a low-angle close-up of someone shielding his or her eyes and then gazing skywards, backed by appropriate take-off sound effects or music, would ease viewers into the skies and prepare them for the following air-to-air and air-to-ground material. Such a shot could obviously be taken anywhere at any time.

The representation of a particular event can often be greatly improved by combining shots actually taken on different occasions. Remember how we decided to unite those two rather forlorn short sequences devoted to the lake trip and the steamer itself? If we *intercut* these so that a shot of the vessel leaving the quay is followed by one taken on board — say, the family on deck looking at the passing scenery — then continue with another long shot of the steamer crossing the lake, and so on, we create a much improved all-round impression of the experience and one that has structure and a feeling of cohesion.

The same technique can also be usefully applied to the climactic excursion, dealing with the trip up the mountain by cable car. Probably there are assorted shots of the cable railway system itself (general views of the suspended cars at various stages of ascent and closer views of passengers boarding the vehicles at the terminal) as well as those showing views *from* the car as the family made the exciting journey to the summit.

If we intercut these separate sets of shots in an intelligent way, the whole sequence will be bound together and most of the jump-cut problems will disappear at the same time. If a few reaction close-ups of the family members as they survey the scene can be included here, all the better.

### Further refinements

Although our planned assemblage of shots has given the original material logical progression and an overall shape, the editing process is still not complete. The pace of the movie is likely to be much too slow, some pieces of action may be duplicated — in short, it'll have a cobbled-together look. The next stage, then, is to smarten up this untidy conception into a smoother and refined form.

Because video editing is a re-recording process, there can be no subsequent transposing of shots or trimming away a few frames after our movie has been assemble edited. Every time we switch the position of a shot or shorten it, *everything* coming after the change has to be dubbed again from scratch — which if it happens with shot No 2 would mean just about the whole video! In view of this, we need to work out and then refine the rough version of the movie on paper before doing any actual editing at all.

When embarking on this refining operation, the first job is to get the pace right for each individual sequence. To a certain extent, pace is governed by the speed of the action within shots, but it can also be further modified by the editing. The cutting tempo can have a lot to do with putting over the atmosphere of a scene, and choosing the wrong one can completely wreck the effect you are trying to achieve.

The rate at which shot follows shot must reflect the mood of the sequence. For instance, the leisurely feeling of our lake steamer trip would be characterised by a fairly slow cutting rhythm (which, of course, will also give the viewer time to take in the passing scenery), while the hustle and bustle of the opening airport sequence would need far brisker editing.

Perhaps it hardly needs saying that this adjusting of tempo at the editing stage can only be done in one direction — that of speeding up the pace by shortening the length of shots. If you discover that a particular take should have lasted for twice as long as you actually pressed the camcorder release, well that's too bad! All of which proves that it always pays to think before you shoot.

There are a few general points with regard to the trimming of shots quite apart from considerations of tempo. First, viewers will normally expect to see a prominently featured action completed within a shot, so their expectation should be anticipated and fulfilled —

that is, unless the action is to be continued in the next shot (see cutting on action, Chapter 8).

Other points to watch out for are those moments at the start and/or end of a shot before characters enter the picture or after they've moved out of it; leaving in such 'dead' frames needlessly slows down the pace. (Naturally, if there's some positive reason for holding the shot, then it's all right to do so.) Professional movie makers are especially alive to this business of dead frames, which is one of the main reasons why their work usually looks so much slicker than that of amateurs.

I've been discussing editing in relation to putting together unplanned holiday material, but a similar approach can be used when tackling a scripted movie. Naturally, in this case the actual order of the shots will be determined by the script, but even then, the editor is still in a position to exercise considerable artistic control over the finished product. Editing is a dramatic device in its own right, and besides being able to affect the mood of a sequence (as we've already seen), the editor can use tempo and rhythm to evoke an emotional response in the viewer.

It's hardly to be expected that such a complex and fundamental subject as editing can be covered exhaustively in a single chapter. However, if you've absorbed the basic principles I have been able to deal with, I can promise that, when put into practice, they will improve the quality of your video movies by several hundred per cent.

## THE THREE STAGES OF EDITING

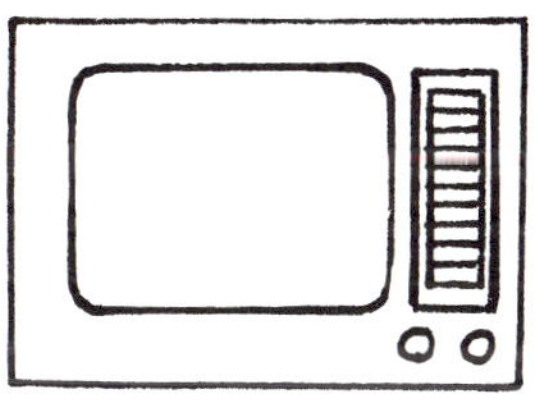

**1** Before any editing is attempted, the material must be viewed and all shots listed with their tape counter numbers. Any technically substandard scenes should be noted for rejection.

**2** Using these shot lists, which can be cut up and stuck together in the required new order, the movie is edited on paper. If any shots need to be trimmed, details should be noted at this stage.

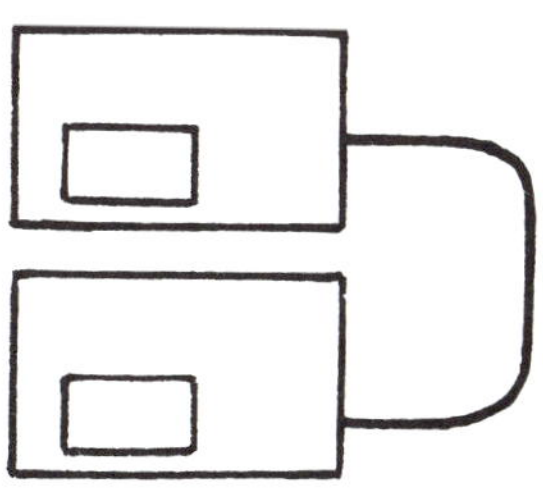

**3** The final version of the video can now be prepared in accordance with the paper edit by transferring the wanted shot material, one shot at a time, to a new tape. Two linked VCRs are used to do this, or a camcorder can take the place of the playback VCR if it has the necessary facilities.

# 11

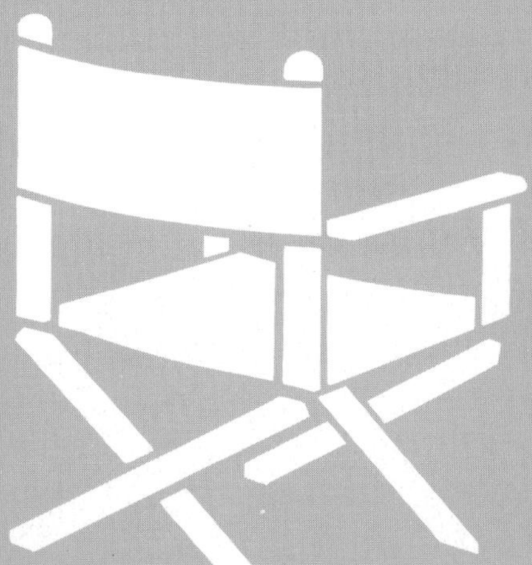

# Chapter 12
# Sound: Equipment and Technique

Some readers may be wondering why I've gone through eleven chapters and haven't yet dealt with video sound. This has been a deliberate policy; movies, in my view, are primarily a visual medium, so I felt it best to concentrate on the basics of picture making before introducing readers to the whole new area that's opened up by the soundtrack. But it's time the balance was redressed and we'll now turn our attention to the audio side of video making.

People are usually surprised to learn that, in the professional world, most of the sound is put on a movie after it has been shot. Those sinister footsteps in a dark alley, the gurgling stream and birdsong in the country, the crashing thunder and howling wind of a storm and, of course, the music — all are the result of skilful post-production work in the dubbing theatre. In fact, about the only sound that professionals record 'live' — that is, while the visuals are actually being shot — is dialogue, and even this is sometimes only used as a guide track for a better recording under studio conditions if the original was done out of doors.

Needless to say, few amateurs have access to the full resources of a dubbing theatre, but it's amazing what can be done with a couple of tape recorders, a simple mixer — and some creative ingenuity. I certainly don't want to scare the beginner by making the whole thing seem hopelessly complicated; I'm just trying to put over the point that there is more to making a soundtrack than going out and recording through the camcorder's built-in microphone. If you do this and hope to come home with a professional-sounding production, I'm afraid you're doomed to disappointment. However, as live-sound shooting is what folk tend to think about first, we'll start by dealing with this.

By its nature, video is essentially a *single-system* method of making sound movies, that is to say, the sound is recorded on the same tape as the picture at the time of shooting. This has its advantages and snags as we shall see.

### Microphone placement

The most important point in connection with shooting live sound is the matter of microphone placement. The object is always to get the greatest proportion of wanted sound, and to achieve this we have to place the mike as near as possible to the sound source.

Now, as all camcorders have microphones built into them, you may be wondering how you can follow this advice. Well, of course you can't unless you're shooting a close-up. These on-camera mikes are fine for recording general atmospheric sound — say, fairground noise — but are not so good at picking up speech if the speaker is more than two or three metres away. In such a case a separate mike on a lead should always be used, so it's wise, when choosing a camera, to make sure it has a socket for attaching one of these.

At this stage I should point out that there are different types of microphones. The *omnidirectional* mike is equally sensitive to sound coming from any direction and is the sort that is

*An omnidirectional microphone (left) is equally sensitive to sound waves coming from any direction, while the cardioid unidirectional microphone (right) is most sensitive to those from the front and is virtually 'deaf' to anything coming from the rear.*

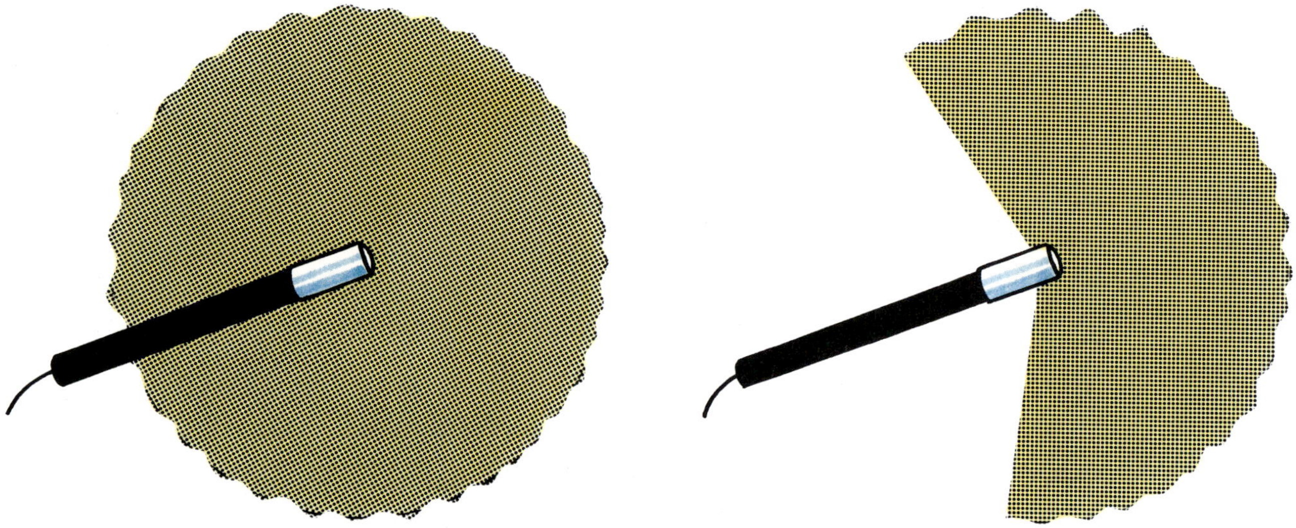

often fitted to a camcorder. The *cardioid* mike, on the other hand, is *unidirectional* and picks up sound coming mainly from the front, its sensitivity falling off rapidly to the sides until it 'hears' almost nothing from the rear. This characteristic is obviously very useful for helping us to record the sound we want because if we turn our cardioid mike towards the sound source, unwanted noises from other directions will be largely eliminated. There are also highly directional *gun* microphones that can be aimed at the sound source from greater distances, but these are rather specialist items of the sort often used by professional news teams.

Another factor that will affect the quality of our sound is the acoustics of the surroundings in which we are recording it. Generally speaking, when recording indoors we don't want too much reverberation — that is, sound waves bouncing off hard surfaces — because this will give an unnatural echo-like effect. This can be minimised to a certain extent by placing the microphone as close as possible to the subject so that the direct sound waves will drown out the reflected ones. But it's best to make recordings in rooms with carpeted floors, curtains and plenty of soft furnishings as these materials all act as sound absorbers.

### Automatic gain control

Amateur video recorders and camcorders have a feature called *automatic gain control* (AGC), the object of which is to keep the sound signal at an optimum level for a fully-modulated and undistorted recording. The intention is admirable but, like automatic focusing, the device is something of a mixed blessing because it's unable to discriminate between wanted and unwanted sound. It only 'knows' whether the sound coming in is creating a signal above or below the predetermined level. If it thinks things are too quiet, it boosts the signal until this level is reached. The result is that background sound which we'd normally wish to remain at a low level, will be brought up to near thunderous proportions if there's no other main foreground sound to keep the system 'happy'.

Perhaps we're shooting in the country and want the birdsong to record as a discreet chirruping in the background, but the AGC will amplify what it regards as a too-low signal until it sound as though we're recording in an aviary. Then if a character near to the mike suddenly speaks, the first syllable will drastically overload the system until the AGC throttles back the signal, with the result that a nasty blast of distortion is put on the track.

There's another snag. To avoid pumping up the signal the instant someone stops speaking and there's a moment's silence, the AGC's 'recovery' time after a peak of sound is relatively slow. This means that if, for example, a character were to give a loud scream in the middle of a dialogue passage at normal volume, the following speech would be under-recorded until the AGC got over its 'fright' and brought the signal level back to where it was.

This is very frustrating for those who would like to have realistic relative levels of sound. However, it *is* possible to 'beat the system' by inserting the equivalent of a volume control (potentiometer) into the lead between the microphone and the camcorder/recorder. The object of using this device is to limit the overall signal level so that it will never be strong enough to exceed the AGC's pre-set threshold.

This limiting control is set initially by adjusting it against an average type of sound signal — such as speech — while monitoring the system through headphones. Starting at zero (when nothing will be heard), you gradually turn up the control until a point is reached where the sound through the 'phones gets no louder. This is the AGC's threshold and the setting should be marked, because from this point *downward* the recorded signal is susceptible to the manual control we're looking for.

This ploy is obviously not possible where the

mike is already built into the camera, so the best we can do in such a case is stick to the average recording conditions for which the AGC is designed — and I suppose it's fair to add that, nine times out of ten, this can be done without too much cramping of our creative style.

### Avoiding aural 'jump-cuts'

A moment's thought will show that when we're shooting single-system (as with video), sound will only be recorded while the camera is actually running; this can create difficulties with certain types of subjects.

Imagine you're shooting a carnival procession which includes a band. The camera is run for, say, ten seconds to get a shot of the band, then stopped while you change its angle to take a reaction shot of your children's excited faces. But while the camera is stopped the band continues playing and is now half-a-dozen bars further on with the music. Result: a nasty aural 'jump-cut' on the soundtrack.

This highlights one of the weaknesses of single-system and illustrates its unsuitability for subjects that have continuous and progressive sound. In such circumstances we shall get far better results by recording continuously with a separate portable sound recorder (that is, using *double-system*) and dubbing this material on to the video after editing. What may be lost in precise synchronisation (usually not noticeable with this kind of background sound anyway) will be more than made up for by a smooth, uninterrupted track. This brings us to the subject of post-production soundtrack compilation.

### Post-production sound

As I explained at the start of this chapter when talking about professional practice, the object of post-production sound techniques is to add to the soundtrack, under controlled conditions, those elements that can't be recorded conveniently or satisfactorily at the time of shooting. There are various approaches, and which we use will depend on the apparatus at our disposal and the pains we are prepared to take.

I suppose it has to be admitted that the video maker is somewhat limited by the amateur equipment at his disposal when it comes to post-production sound. Existing VCRs have no *sound-on-sound* capability for superimposing one sound signal over another that's already been recorded

*The insertion of a volume control (potentiometer) between microphone and recorder will enable you to 'defeat' the automatic gain control (AGC) to achieve more realistic relative sound levels. This device will limit the overall signal level so that it will never be strong enough to exceed the AGC's pre-set threshold.*

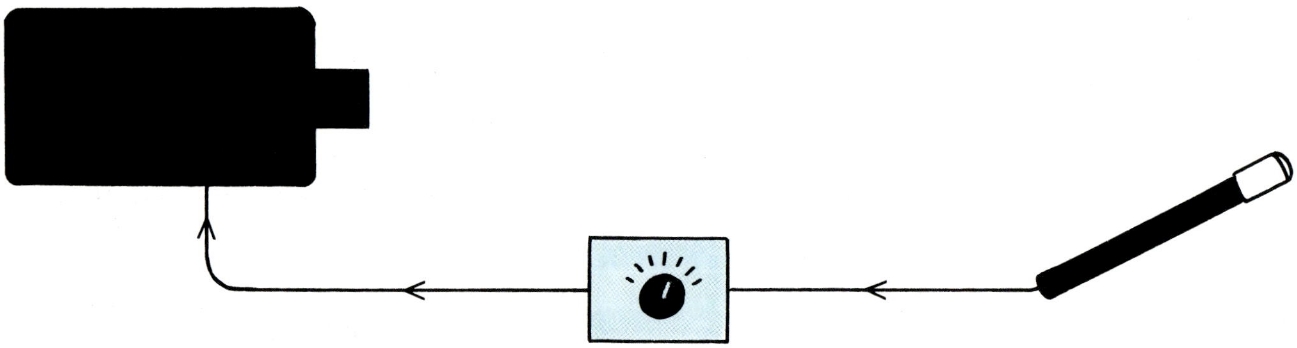

— which would certainly be a very useful facility in soundtrack compilation.

However, many video recorders do have an *audio dub* feature built in which enables a completely new track to replace the original sound of a particular set of visuals. At first sight, this might seem a facility of rather doubtful value, but it can prove very useful, as we shall see.

Although the sound resources offered by present domestic video equipment are limited, it is possible for the serious worker who is prepared to take a little trouble to compile relatively sophisticated soundtracks with the aid of additional apparatus that is readily obtainable.

### Twin-track recording

The simplest way of adding extra sound is by means of twin-track recording, and to do this you need a VCR with a stereo sound facility. When this method is used, the stereo video is first edited into its final form (as described in Chapter 11). This done, the additional sound material is recorded on to one of the stereo tracks of the tape, which means that on replay you will have the original 'live' sound plus the extra post-production recording. The snag about this method is that you may have some initial difficulty judging the correct relative levels of the two soundtracks, and some experimenting may be necessary to get it right.

In most cases it will be desirable for one track to predominate. If, for example, your original live sound is made up of background effects — say, sea, wind and gull cries in a beach scene — and you wish to overlay these with a commentary, then naturally this will need to be recorded at a higher level so that the words come through clearly. On the other hand, if the original sound is a conversation which you want to underline with suitable music, then the second recording will need to be at a lower level than the first.

Because of the problem of balancing the two levels reliably, the twin-track method of adding extra sound is something of a make-shift procedure, but there are more sophisticated ways of doing the job 'properly' so that all the audio ends up in its finally mixed form on one track in the approved professional manner.

### Lift off, mix and put back

Here the technique is to 'lift off' the original sound by transferring it to ¼in tape. The extra sound is then added on another track of this tape and the combined audio material transferred back to the video soundtrack in sync.

This method requires the use of a suitable synchroniser, the function of which is to relate a given length of sound tape to an equivalent number of picture frame pulses on the video tape and keep the sound recorder and VCR in step.

Another technique for transferring video sound in sync works on the principle that, for all practical purposes, two VCRs will run at identical speeds. After editing the programme into its final form, you copy the *edited version* on to a third cassette. The additional sound you wish to add is then prepared on ¼-in tape by a system of cueing that we'll look at later on.

To carry out the sync transfer while adding extra sound you set up the equipment like this: connect the output of the sound tape recorder and the *audio out* of a VCR (call it A) to the input of a sound mixer; feed the output of this mixer into the *audio in* connection of a second VCR (call it B) and connect the output of this VCR to a TV monitor.

The ¼-in tape with the to-be-added sound is placed on the tape recorder, while the *copy* of the edited video and the edited version itself are put on VCRs A and B respectively. Both video tapes are then inched up to the same starting point by means of the frame advance; this is important because the sync will depend upon the accuracy with which this is done.

The tape recorder and the two VCRs are then put on pause before being started together — if you don't have three hands you'll need an assistant for this!

# 12

What's now happening is that the extra sound from the tape recorder and the sync sound on the copied video playing on VCR A are replacing the original sound on the edited video running of VCR B. It will be necessary to use the mixer to balance the two sound signals 'on the run' to get the correct respective levels. This shouldn't prove too difficult, and anyway, you can always have another try if you don't get it right first time.

Just to make it a hundred per cent clear, I must stress that it's only the *sound* from the copied video that we transfer; the picture on VCR B's tape is retained and unaffected by the operation, the VCR's audio dub control being used to make this new recording. Admittedly, the original video sound will now be third generation — that is, at two removes from the 'live' recording — but this is relatively unimportant, especially if the original was of good quality.

It's worth mentioning that the quality of the re-recorded sound is likely to be improved during transfer if a tone control is placed between the mixer and VCR B; some top cut will reduce tape hiss. Also, the mixer can be used to control the overall sound level to 'defeat' the automatic gain control of VCR B, which is desirable for the reasons explained earlier in this chapter.

I should also point out that although the audio tape with the extra sound is started with the video, it probably won't stay in exact sync with the picture for the whole run because the speed of tape recorders isn't as accurately controlled as that of VCRs, and there are other factors such as slippage and tape stretch. All the same you're unlikely to be more than a second or so out either way, and as the extra sound is probably a musical background or commentary, this can be planned on a loose sync basis from the outset.

Having considered the various aspects of adding extra sound to a previously recorded 'live' track, we'll now turn our attention to compiling a completely new soundtrack for a video whose original sound we don't much like and wish to replace.

### The cued tape system

The idea here is to prepare the complete soundtrack on ¼-in tape, and then when we've got everything right, to transfer it to the video's audio track, using the VCR's audio dub facility.

The basis of the technique is the preparation of a pilot tape which is closely cued to the visuals. Once this guide track has been prepared, the sound can be built up stage by stage without constantly having to re-run the video. It also avoids the somewhat nerve-racking necessity of watching the screen when finally recording the commentary.

The first step is to make a start mark on the audio tape, which is then lined up over the recorder head. In the case of the video, the first picture frame acts as the start reference which you can find by using the VCR's frame advance.

The tape recorder should have been run for about ten minutes beforehand so that it will have warmed up and stabilised its speed. If very precise sync is required or the movie is longer than about twenty minutes, a synchroniser of the sort previously mentioned should be used to make sure the VCR and recorder don't drift out of sync.

The drill is now to start the video and tape recorder (which should be in the record mode) together. Watching the picture, you call out a number at each shot change — '2' for the second shot, '3' for the third, and so on to the end of the movie. I must stress that the calling of these cues must be as accurate as possible because the final sync will depend on this. It's not a bad idea to have a dummy run so that you can anticipate where the shot changes will come.

Once the sound tape is cued, we re-run the video and methodically prepare a numbered list of the shots with a brief description of each. I call this the *skeleton script*. This script and the cued tape are all we need to compile our soundtrack; the former is our guide to what sound is required, and the latter defines its duration and placing.

For example, if we feel the first four shots

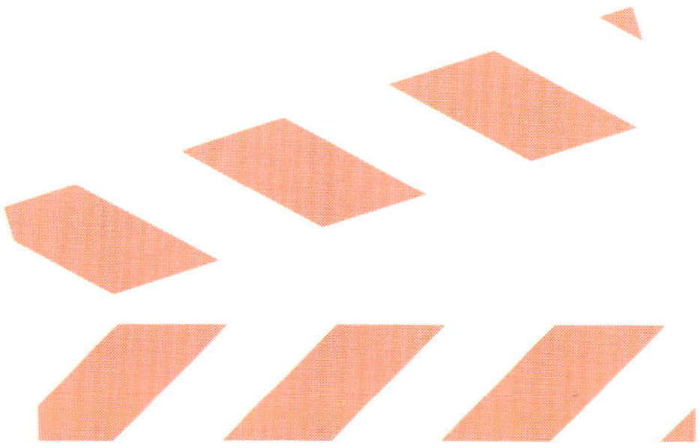

provide the basis for one section of commentary, we bracket them together and label the group 'A'. Returning to our recorder, we find the point on the tape where we called shot No 1 and, using a yellow chinagraph pencil or small adhesive sticker, mark it with a cue 'A'. Using a stopwatch while running the tape, we time the duration from cue 'A' to the end of shot No 4 (the instant before shot No 5 is called). We note this timing against section 'A' on the skeleton script.

The next section of commentary — which might be the following three shots — will be labelled 'B' and timed in the same way. If the next five shots after these are to be accompanied by music only, we still bracket them and cue them as section 'C', but indicate on the script that no commentary is required here.

*The set-up for a mix of 'live' sync and added sound. A copy of the video is run on VCR A, but only the audio content is fed out. The extra sound (on ¼-in tape) is played back on the reel-to-reel recorder. The combined signals, after mixing and the application of tone control (to cut tape hiss), are fed into VCR B where they replace the original video sound by means of the audio dub facility. The monitor is used to enable both video tapes to be inched up to the same frame for a sync start.*

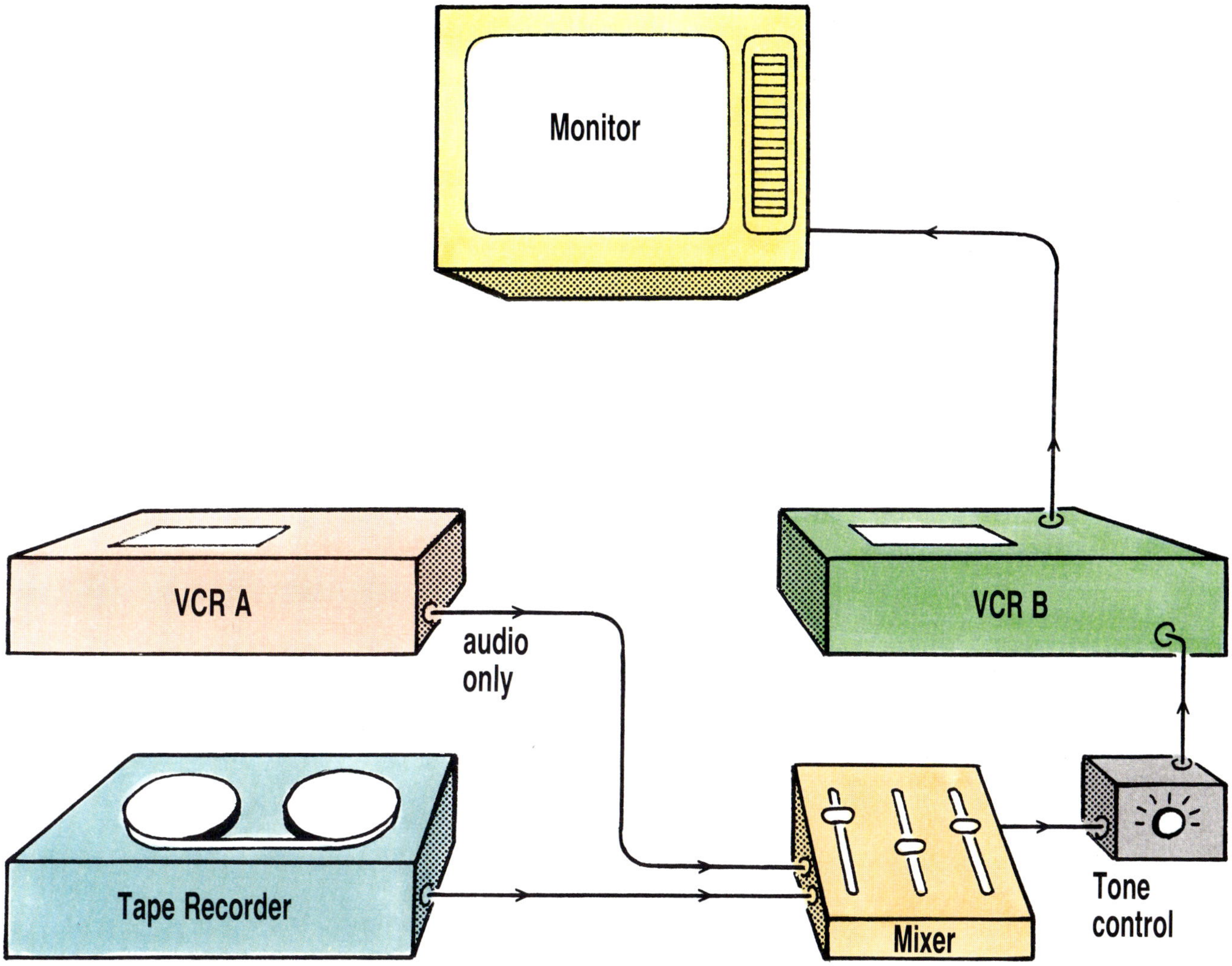

# 12

The whole soundtrack is plotted in this way and the beauty of the system is that it can be done without further reference to the video.

The recording is eventually done on the pilot tape itself, the cues for sections of commentary, music or background effects being taken from the visual A-B-C marks that we've been putting on during the timing process. As it's obviously a little difficult reading a commentary while watching for cue marks on a moving tape, it's best to record the complete script with just a slight break between sections on a cassette tape. If the cassette recorder is then connected to the ¼-in machine, the commentary can be dropped in at the appropriate points as the cues come up on the master tape.

Similarly, sections of music or effects can be built up on the other parallel track of the ¼-in tape using the appropriate cues to place them. As the music/effects *are* dealt with separately from the commentary, each can be recorded at whatever level is found to be most suitable, and if either is fluffed in the process, it can be done again without affecting the other.

When the soundtrack has been completed to our satisfaction, it can be transferred to the video's audio track in one operation by playing both tracks of the ¼-in tape together and feeding the combined signal into the VCR, using the audio dub control.

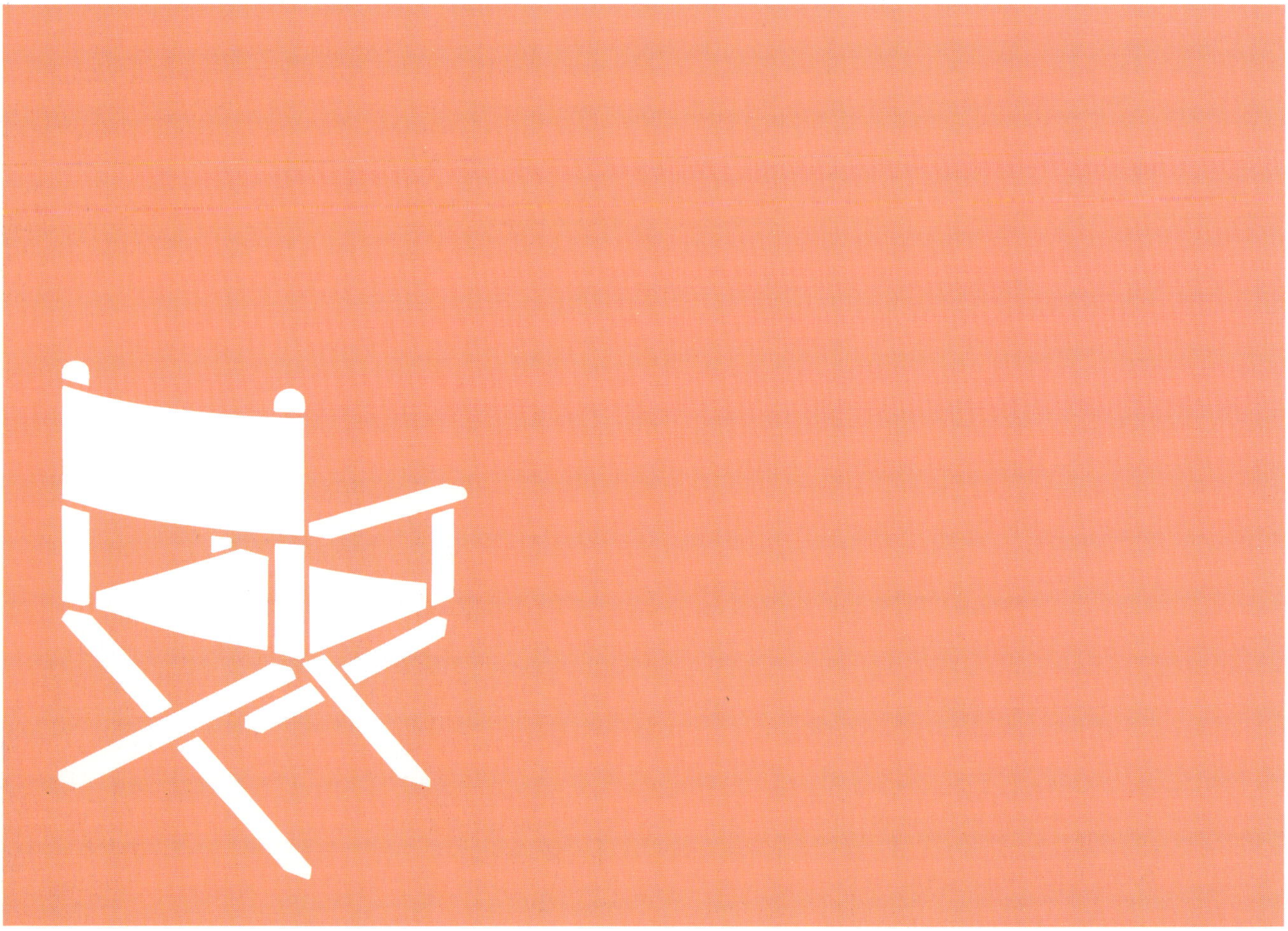

# Chapter 13
# Sound: Choosing and using words and music

In the previous chapter we considered the mechanics of adding extra sound to 'live' sound and even completely replacing it by post-production techniques. So it's appropriate, now, to consider the artistic side of the process, and take a look at the business of commentary writing and choosing and using music.

At first sight, it might seem that creating a commentary for a personal holiday movie ought to be the simplest thing in the world. After all, we were there at the time, know what happened so surely just have to 'tell it like it was'. But things aren't quite that simple, and there are a number of pitfalls that need to be pointed out to beginners.

A basic point to be made is that a commentary should never be a stream of non-stop chatter, delivered *ad lib* while the video is running; the impromptu commentary with its hesitations and

# 13

corrections sounds just about as amateurish (in the worst sense) as you can get. The key to success in this, as in most aspects of movie making, is advance planning, so our thoughts should be worked out and refined on paper from the start.

Naturally at the beginning of any movie viewers need to be put in the picture, so with a holiday record we should open our commentary by sketching in such basic facts as where we went, the time of year and possibly what prompted us to go there. As these are the sort of questions that are likely to arise in viewers' minds, we immediately put them in a receptive mood by supplying the appropriate answers.

You'll recall that in Chapter 12 I talked about the skeleton script — a numbered list of the movie's shots with a brief description of each — and how shots were grouped into convenient sections which were then timed. Before attempting to write our commentary we need to refer to this script and its sections to see what needs to be said and how long we have to say it. From the timing point of view, if we work on a basis of three words a second this should be about right for most subjects. Obviously it isn't an inflexible rule because solemn material might demand a slower rate of delivery and so fewer words per second would be suitable.

The most important point to grasp about the art of commentary writing is that the words are there to *complement the pictures* — that is to say, fill in information that the viewer needs to know but is unable to gain from the images alone. They mustn't just describe what can be clearly seen on the screen. This fundamental rule applies equally whether the movie is a record of your winter

*On the left, a portion of the 'skeleton' script consisting of a detailed list of the shots in the edited movie. These are grouped into sections which are given a cue letter and timed. The commentary script, right, can then be written section by section without further reference to the video itself by using the skeleton script as a guide to the action and screen time.*

| Shot | PICTURE | SECTION |
|---|---|---|
| 20 | LS — General view of Positano showing houses on hillside sloping down to sea | G 12 sec |
| 21 | MLS — Family walking along street having a look around | |
| 22 | MS — Artist sitting at easel, painting | H 17 sec |
| 23 | LS — Showing artist's view | |
| 24 | CU — Canvas, showing artist's interpretation of scene | |

| SECTION | COMMENTARY |
|---|---|
| G 12 sec | Positano — situated between Amalfi and Sorrento — was our next port of call, and we lost no time in getting out of the car to explore this fascinating little town. |
| H 17 sec | It's small wonder that the place holds such an attraction for artists when you have these colourful houses on every side — and such wonderful light to paint them by. |

sports holiday or a sombre documentary about the dangers of pollution from nuclear power stations.

### No need to be 'literary'

Some people, who possibly doubt their writing ability, tend to quail at the thought of being asked to put pen to paper. However, their fear is quite unnecessary because 'classy' prose in the literary sense is not what's required. For instance, what we don't want is this sort of thing:

> *As we made our reverential approach towards the Sistine Chapel, it was as though every step shook from our world-weary feet a little more of the Twentieth Century's dreary dust. We fancied we could almost breathe an enigmatic blend of art and history that seemed to hang in the air like a rare and subtle fragrance, causing our expectant aesthetic senses to tingle with a new heightened awareness . . .*

This is obviously an extreme example of the overwritten, self-consciously 'literary' style, but it does illustrate the dangers of trying to sound clever. Also such writing is hopelessly lacking in solid information. We'd be unlikely to talk in this high-flown way if recounting the holiday to a friend; on the contrary, we'd use everyday language to express our reactions to the occasion. There's no reason to go all arty because we're putting the thoughts on paper.

An improved version of this section of commentary might go:

> *During the second week of our stay, we went on a day excursion by coach to Rome, which included a tour of the Vatican. As Sandra is a bit of an art buff, she was naturally delighted that we had a chance to see the Sistine Chapel, famed for its painted ceiling by Michelangelo.*

This, I think you'll agree, is much better. Gone is the purple prose verbosity with its rather precious imagery, and in its place we have some genuinely useful information: viewers are told when and how the visit was made, the fact that Sandra is an art-lover and (just in case they didn't already know) why the Sistine Chapel is famous.

If pretentiousness is one possible vice in commentary writing, an over-colloquial, 'matey' style is another that's equally to be avoided. It often takes the form of banal patter laced with feeble attempts at humour:

> *And I've got a feeling that when we turn this next bend we shall see the Leaning Tower of Pisa — actually, it was about here that Sandra started to feel sick. Ah yes — here it is coming up now — the Tower, that is, not Sandra's lunch. My word! Looks a bit dodgy, doesn't it? In fact, when we first saw that tilt we nearly crossed it off our list! Tilt, list — get it?*

When writing a commentary we should strive to strike a reasonable balance between what sounds arty and contrived, and what comes out as mere garrulous chat. In other words, the aim should be to write as we generally speak, without affecting airs and graces, but that isn't to say that we shouldn't also smooth off some of the rougher edges of everyday speech in the process.

Another point to bear in mind is that the commentary should not go rattling on and on without a break. It can be tiring for viewers if their visual and aural attention is engaged continuously; therefore try to arrange things so that each sense in turn is given a rest from close concentration. This means that when the picture is making an important point, we should ease up on the words. Similarly, when the commentary is imparting vital information, we shouldn't accompany this with a display of visual fireworks.

### Voice levels

Having written our commentary we now have to record it, but before doing so, it's a good idea to do a bit of experimenting with different

# 13

microphone placements and voice levels. If one speaks at a low level with the mike close to the mouth, the voice when played back will sound intimate and confidential. On the other hand, if the mike is placed further away and the speaker adopts a more forceful tone, the result will be strident and authoritative.

Both approaches can be equally useful in the right context and which you use will obviously depend on the nature of the subject. The more muted delivery would suit, say, a religious procession, while the punchy style would be just right for a sports event.

It perhaps hardly needs saying that the reading of the commentary should be rehearsed before the final recording is made. If this rehearsal reveals that some phrase is difficult to say, then the offending wording should be recast so that it trips more easily off the tongue.

Regarding the actual delivery, the rule should always be to sound as natural as possible. Whatever you do, don't try and copy the mannerism of professional commentators you've heard, and if you have a regional accent, don't attempt to disguise it; your natural voice will give the movie a warmer, friendlier touch.

A common fault to guard against is letting your voice drop at the end of every sentence, which is a certain way to create the kind of monotony likely to put viewers to sleep.

Unless your movie is fairly short, it's probably asking too much of an amateur commentator to tackle the recording of the whole script in one continuous take. Even if assistance is to hand on the technical side of the operation, recording sessions often prove to be rather tense occasions. Best, then, to have a break or two so that tension can be relaxed over a cup of coffee — but be sure that all the recording conditions are the same when you start again, otherwise there may be a disturbing change in the acoustic effect half way through the movie.

The physical comfort of the commentator is also something that must be considered; he or she should be able to feel relaxed and, if recording to picture (as opposed to cued tape), have a clear view of the screen. However, relaxed comfort doesn't mean being slumped in an armchair; in fact, this is the worst possible posture because it tends to hunch the shoulders and restrict the breathing. Far better to sit up to a desk or table at a comfortable height in relation to the seat.

The script should be clearly typed in double spacing on thick paper so that there's no rustling when a page is turned. Ideally, there should be no turn-overs in the middle of a commentary section. Obviously, the script should be adequately lit for easy reading.

## Music

The first thing that needs to be said here is that amateurs can't use any music they like and stay within the law. There is that little matter of copyright which makes it illegal to re-record commercial discs or tapes, even though the resulting soundtracks are only to be played back in private.

But fortunately there is an answer to this. You can make use of mood music that has been specially produced for amateurs and has no copyright restrictions when re-recorded in this context. This can often be obtained through specialist audio-visual suppliers.

You could, of course, play and record your own music, or get a musician friend to do it for you. Domestic electronic organs are useful for this purpose.

Knowing where to get music is one thing, but understanding how to use it is quite another. Naturally, personal taste will tend to determine precisely what music we use to back a particular sequence. But having said that, the piece chosen must have a basic suitability to the action. It wouldn't be a good idea to accompany scenes of little Johnny playing in the sandpit with something like Grieg's *In the Hall of the Mountain King*. Quite apart from the menacing character of this

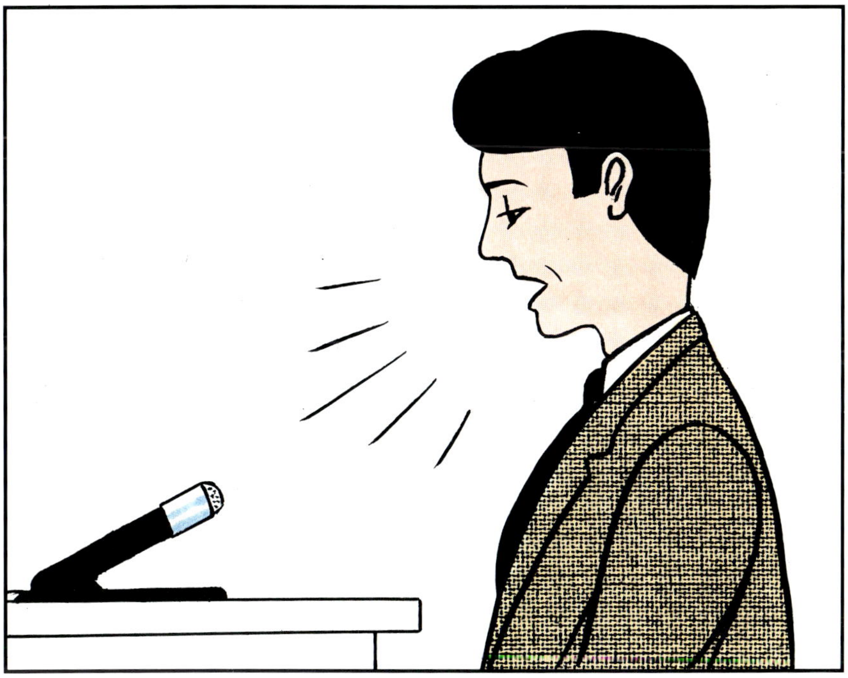

*If the commentator sits well back from the microphone and uses a forceful style of delivery, the recorded commentary will sound strident and authoritative — apppropriate for lively subjects, such as sports events.*

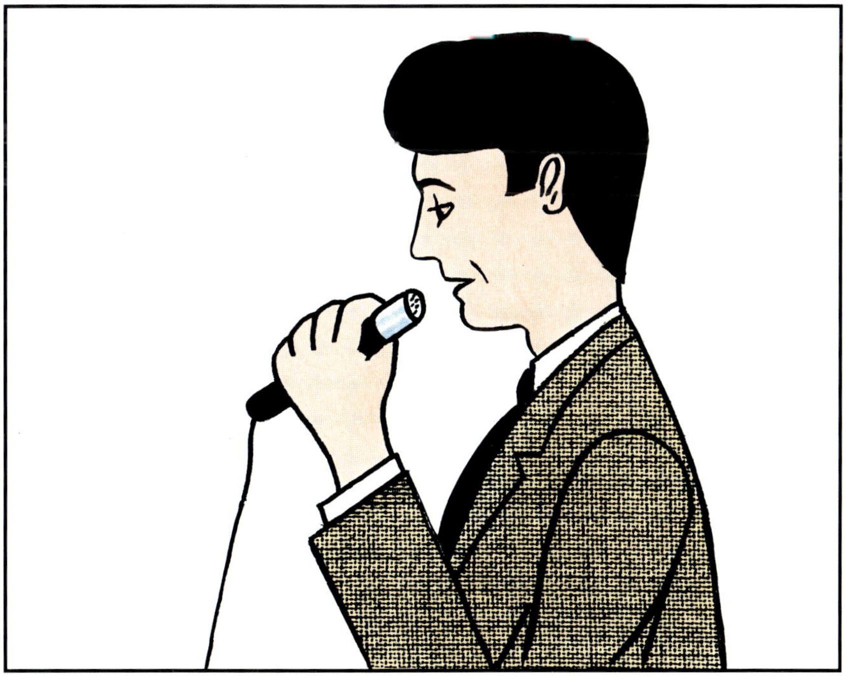

*When the commentator speaks close to the microphone and adopts a low voice-level, the recording will sound intimate and confidential — suitable for pastoral or solemn subjects.*

music being totally at odds with the essentially playful nature of the action, the *scale* of the accompaniment is wrong, the symphonic orchestral resources overwhelming the lighthearted emotional content of the sequence.

So the basic rule is that the style of any music we use must echo the nature, mood and pace of the visuals. Maybe you feel that in saying this I'm merely providing a blinding glimpse of the obvious, but during a fairly long career of amateur movie watching I've seen and heard enough grotesquely inappropriate 'matchings' of music and images to know that the point is far from being universally understood.

Even if they *are* well matched in spirit to the action, it's generally a mistake to use familiar popular classics as an accompaniment. The point is that these usually have their own special associations for individual viewers which are likely to undermine rather than enhance the mood you're trying to create for *your* video.

Another common abuse of music occurs because it tends to get 'used by the yard', by which I mean that defenceless melodies are chopped about and cobbled together in a way that's cruel to both the original composer and the viewer. The question of music editing is a whole subject in itself, but here is a tip for avoiding the most blatant of amateur 'sins'.

What should *never* be done is ruthlessly cut off one track in mid phrase (because that sequence of the movie has ended) and come crashing in with another piece of music of different character and possibly in a different key. The jarring effect is slightly softened if the first track is faded out and the second faded in, but the result still sounds pretty amateurish, and has the viewer feeling left 'up in the air' because the original music is incomplete.

### A proper conclusion

The trick here is to arrange things so that the first music track comes to its proper conclusion at the end of the picture sequence. This can easily be done if the music chosen is longer than the corresponding section of the movie.

You first time the sequence and find that it's, say, 4min 20sec, then time the *total* length of the music track, which might be 5min 30sec. (The music should have been transferred to tape at this stage.) Next, deduct the sequence timing from the music timing to find the difference. Now rewind the music tape to its beginning and run it for the duration of the timing difference — in this case 1min 10sec.

Stop the tape smartly at the end of the run, and put a cut mark on it at this point. Now if you start the tape and fade in at this mark when making the transfer to your soundtrack, the music will accompany the sequence for its duration and end precisely with it as required.

If the piece of music you particularly want to use is shorter than the movie sequence, the thing to do is record it twice and then use the technique described above with the total recording. This will often work if its conclusion (which will now also be heard in the middle of the sequence) doesn't sound too dramatically final.

Of course, the method I've just described is really a substitute for proper music editing. But if this rather specialised pursuit is not for you, the 'making-ends-meet' technique is the next best thing, and will add smoothness and finesse to your soundtrack to a quite surprising degree.

# Chapter 14
# Titling

If our videos are to have anything approaching a professional look we must *title* them effectively. Doing this should not be regarded as some sort of chore, but as an interesting exercise which is all part of movie making fun. And in case anyone is getting worried, let me begin with the assurance that you don't necessarily need great artistic talent to achieve success.

Every movie needs at least two titles: one at the start — the *Main Title* — to announce what it's called, and another at the finish — the *End Credit* — to signal when it's all over. The design of the Main Title is certainly the most important, for besides performing the practical function of telling viewers what it is they're watching, it should reflect the character of the movie and so set the mood for what is to follow. The Main Title is to a movie what the dust jacket is to a book — an arouser of interest and expectation.

Regarding End Credits, we know a professional movie is over when that long and all-embracing list of technicians and other helpers starts to creep up the screen. Some amateurs like to copy this practice with results that are often, in the context, pretentious and a bit silly. Let's face it; most of us are making personal, family record-type productions and it will look rather absurd if these are festooned with credits such as: 'Photographed by John Smith', 'Edited by John Smith' and 'Directed by John Smith'. If John Smith *is* the only begetter of the movie, then what's wrong with a single credit which proclaims: 'Made by John Smith'?

On the other hand, in a group production where a number of individuals have made a special contribution, it's only right and proper that they should receive a screen credit.

Another category of title is the *subtitle*. This may appear at any point in a movie to give the viewer information such as time and place. However, as the commentary can usually supply these details, subtitles are used less frequently nowadays.

Before moving on to consider how titles are actually prepared, there's a basic technical point that needs to be dealt with at the outset. Because with video there's no physical cutting and joining of tape, the opening titles have to be shot first, which is not always convenient, or space has to be left for them to be dropped in later by insert editing (see Chapter 11). This means we have to anticipate their duration and run off the appropriate length of tape before any scenes of the movie are shot. Of course this only applies where we intend to edit in camera and show the original tape; in the case of an edited copy, it would naturally be possible to incorporate the titles when assemble editing the main material.

### Made-on-location titles

As the term implies, these are titles produced on location at the time of shooting, making use of any suitable props or materials that are to hand. They can take a wide variety of forms, and have the advantage of making it easy to shoot them in sequence.

The commonest example of such a title is the one that's lettered on the beach — a natural for a holiday movie. Here the wording is written with a stick on damp sand, or made up from pebbles and small shells. An elaboration of this idea — if you don't mind going to a little extra bother — is to arrange for the title to be prepared ahead of an advancing tide. Then, by timing the shot in

*A common fault in amateur titles is for the lettering to be too large and the typeface unsuited to the subject. A more pleasing effect is achieved with smaller lettering and, in this instance, a less formal style of type.*

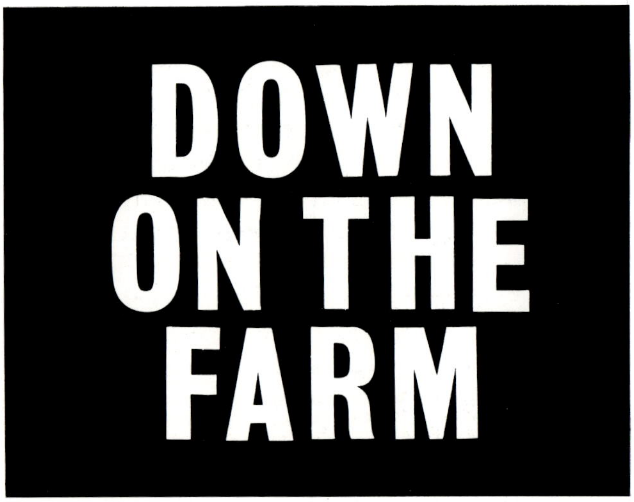


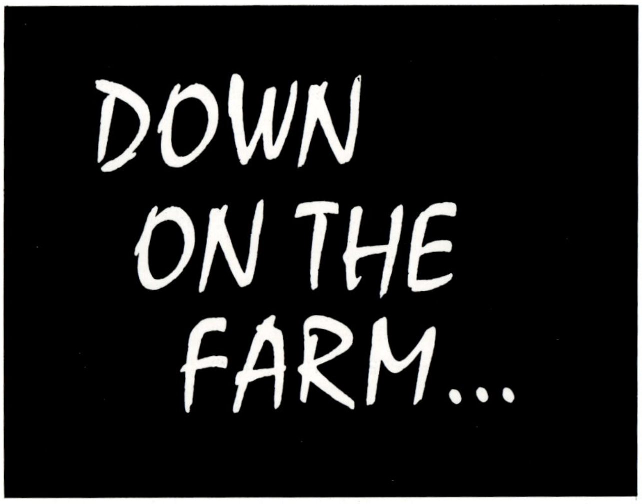

relation to the approaching water, you can have the wording washed away once the viewer has read it — an appropriate trick for an End title.

Obviously, there are many possibilities with on-location titling: titles on banners for 'demo' or 'protest march' movies, writing on a blackboard for a documentary about education, lettering made up from twig arrangements for a woodland fantasy, and so on.

The beauty of titling this way is that it needs no special equipment (except a tripod in certain cases) or practical artistic ability beyond a basic sense of design — plus, of course, a touch of imagination. But perhaps even more important, such titles establish the mood of the movie by reflecting its nature in their graphic style.

If you possess one of the more elaborate cameras incorporating a caption generator, you have the option of superimposing short subtitles (place names, dates, etc) at the time of shooting. Electronically produced lettering is obviously fine for this purpose, but I feel it's a little too clinical and characterless to make effective main titles — unless, perhaps, the movie happened to be on some scientific or 'high tec' subject when the style would be appropriate.

## Artwork titles

As most video cameras now have electronic viewfinders which show the precise field of view being covered, title cards can be lined up visually and shot with the camera mounted on a normal tripod, provided a few basic precautions are taken. The title must be firmly supported and held at right-angles to the camera lens' axis — it can be fixed to a wall or laid on the floor with the camera pointing straight down.

The artwork must be evenly illuminated. Although natural light will do at a pinch, artificial illumination is preferable as it's more consistent and controllable — but make sure the camera is correctly adjusted for whichever type of lighting you are using. Purists would say that proper photographic lights should be used, but perfectly acceptable results can be obtained from two ordinary domestic 150w lamps in reflectors set on either side of the card at a distance of about 15in from its centre.

As titling involves working relatively close to the subject with consequently limited depth of field, it goes without saying that focus must be correct for the artwork-to-camera distance. If the camera's own lens won't focus down to the required distance, then a supplementary lens may be added as described in Chapter 3.

To recap briefly, a +2 dioptre supplementary lens will allow you to work at a distance of 20in (measured from the card to the front of the supplementary itself), which is convenient for a field size of about 8 × 6in. The precise area taken in will, of course, be determined by the camera's zoom setting.

## Methods of lettering

For those of us who are not artistically gifted, there are kits available to help with making our titles. These contain sets of letters, numbers and basic punctuation marks which can be arrangement over any suitable background. The characters are usually made of plastic, while some brands are magnetic for use with metal background plates.

Kits of this sort are easy to use and produce very attractive titles for just about any type of movie. They can, of course, be used again and again. However, because they are essentially neutral in style, they can hardly be expected to add that atmospheric 'something' which is a bonus of the specially lettered title — although a shrewd choice of background can go some way to making up for this.

Those wanting to give their titles a shade more individual 'character' can make use of rub-down transfer letters. These can be readily obtained in a wide range of sizes and faces, and there's little problem in finding a style to suit any kind of movie. It's true that transfer letters need a bit more artistic skill and judgement to use than

# 14

movable characters, but the technique is not difficult to master with a little practice and patience.

### Basics of design

When it comes to design, the fundamental requirement is that the title must be easy to read; everything else is of secondary importance to this so avoid any temptation to be over-clever or 'arty'.

There are three main factors governing readability: typeface (the style of the lettering), size and spacing. There is also the related matter of the background over which the wording is set. A common mistake is that the lettering is used too large. When this happens, it gives the unsatisfactory impression that the title is spilling out of the screen — an effect sometimes seen (because of the picture cut-off) when films of the Thirties and Forties are shown on television. A good margin of 'air' — at least 10 per cent of the picture height/width — should be allowed around the lettering 'block' comprising the title.

A title containing relatively few words — such as a short Main Title — can safely be set in capital (upper case) letters, provided that the typeface selected is suitable for using them in combination. Some faces aren't; for example, you should never set an all-caps title in Old English (the style once popular for newspaper mastheads) as the result is almost unreadable.

Longer titles, such as credits or explanatory subtitles, should be set in small (lower case) letters with initial capitals where appropriate, as these are easier to read when the text makes more than a couple of lines.

A title's impact can be made or marred by the way the wording is laid out within the picture area. It's a mistake to think that the lettering must always be set bang in the middle of the frame; this is not necessarily wrong, but often a more pleasing effect can be achieved by aligning — or ranging, as it's called — the wording to the left or right.

With a one-line title, it's best to place it slightly above the horizontal centre line of the frame because if set exactly on it, a curious optical illusion causes the lettering to look as though it's actually lower and slipping towards the bottom of the screen. If you want to have the line towards

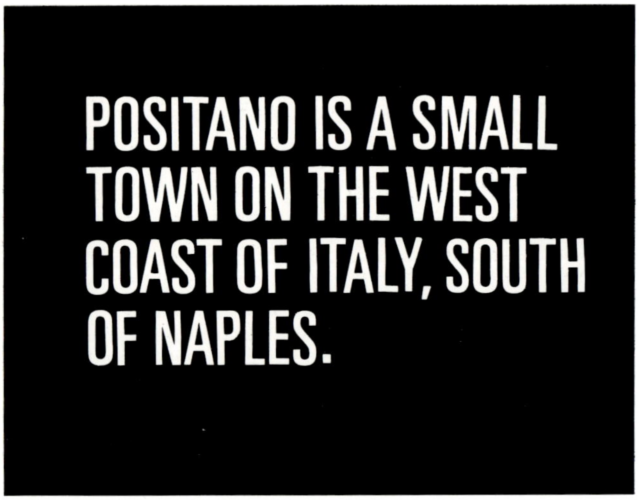


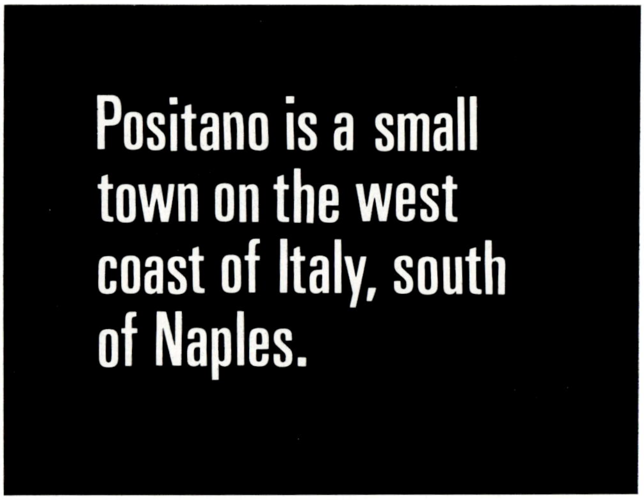


*Captions which run to several lines should not be set in capital (upper case) letters, as these are not easy to read in a 'block'. A more readable result is obtained by the use of small (lower case) letters and upper case where necessary for place names, etc.*

the bottom of the picture, that's fair enough, but make sure it's low enough (say, between a quarter and third of the way up) so that it looks as though it's meant to be there.

The way you arrange the lettering for any particular title will be influenced by the sort of background it's set against. If it's quite plain — perhaps just a single colour (although this is best avoided because of the likelihood of picture 'noise') — then the lines can be laid out in any way that gives a clear and balanced result. If, however, some sort of pictorial background is being used, then it will be necessary to arrange the wording around its main features to make a total, satisfying design. What must always be avoided is placing small lettering over a 'busy' (that is, cluttered and detailed) area, as this makes for illegibility. And ensure that there is good colour contrast between wording and background; this applies particularly to titles superimposed on live action with a caption generator.

Two final don'ts. Never space out a word or line unnaturally just to make it align with something else; the effect is invariably clumsy and amateurish. Also, never use different

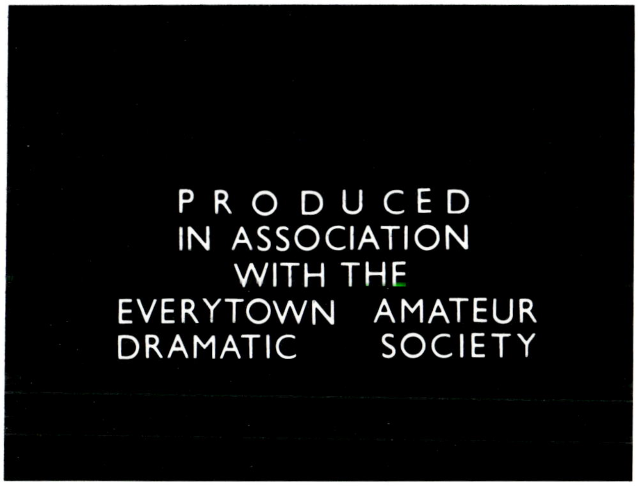


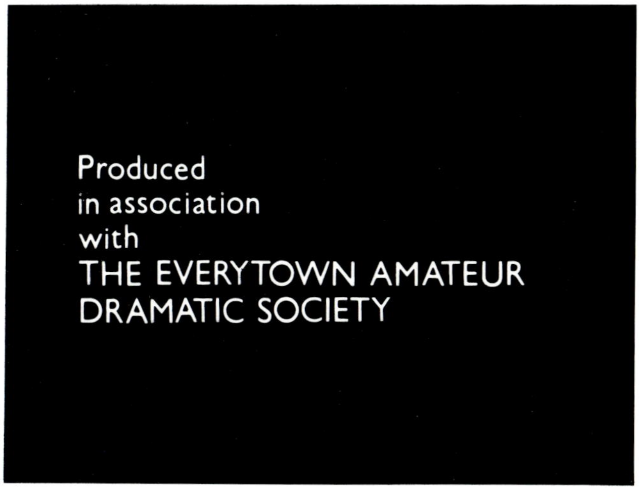


*The top illustration shows some more common faults. The lettering has been set too low in the frame so that it seems to be slipping down the picture. Although there is nothing wrong in principle with a centred design, here it's spoiled by poor spacing — the way the top and bottom lines have been artificially spaced out is particularly ugly. Also more care might have been taken to get the letters in 'PRODUCED' level. Below is a much better layout with the wording ranged left (actually easier to do when transfer letters are used) and set slightly above the horizontal centre of the picture. The mixture of lower case and upper case letters is both more attractive and readable than the all-caps version. A tip on letterspacing: leave most space between letters with adjacent vertical lines (such as IL), less between those with a vertical and round shape (OL), and least between two round shapes (DC). The larger the characters, the more important letterspacing becomes.*

coloured letters to make up a word; this causes the eye to register the individual characters rather than the shape of the word itself which makes it harder to read. However, complete words in contrasting colours can sometimes be effective — especially if one of the 'colours' is black or white.

### Reading time

Now for the age-old question: how long should a title remain on the screen? Obviously, it needs to be held long enough to give viewers time to get its message, but not so long as to outstay its welcome after this purpose has been served.

The snag is that every viewer will not read at the same speed, so the duration must anticipate the reading rate of the slowest. A useful rule of thumb for gauging this is to read the title out loud yourself at normal conversational speed and then read it half through again, the total number of seconds it takes you to do this being the time the title should stay on screen.

Sometimes the Main Title may be an exception; it might very well consist of just one or two words that can be read in an instant and so, according to our rule, should only be held for a couple of seconds. But at the beginning of a movie viewers require a little time to settle down and get into the spirit of things. The Main Title, then, can remain on screen for longer than is strictly necessary to read it — something like seven or eight seconds is generally about right.

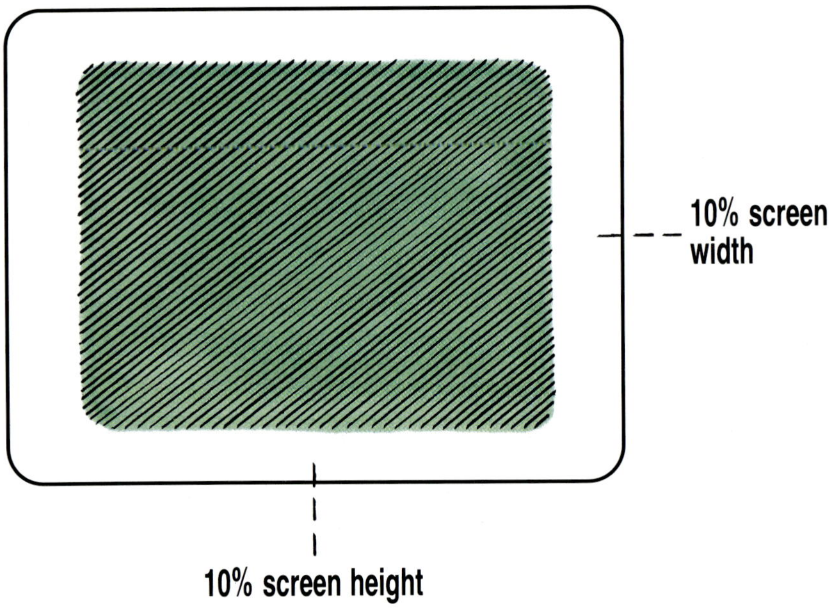


*A title should never completely fill the screen, so leave a margin of at least 10% of the screen width/height around the block of text. If this rule is not followed, the wording looks as if it's in danger of spilling out of the frame, so make sure your title occupies no more than the shaded area indicated.*

# Chapter 15
# Ten Tips On . . .

- Shooting a Wedding
- Making a Holiday Video
- Recording a Sports Event
- Shooting a Children's Party
- Camera Care
- Tape Care

# 15

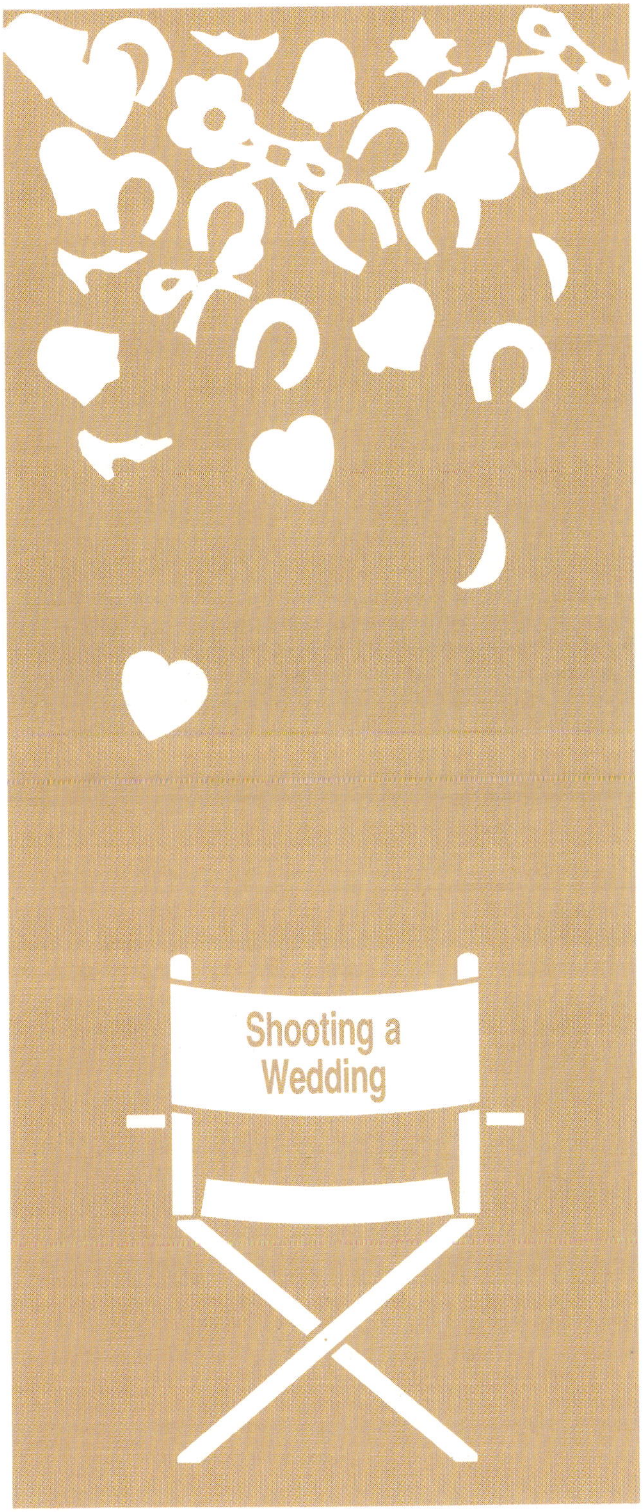


## . . . Shooting a Wedding

**1** The secret of success when making a wedding video is thorough preparation. It's no good leaving every thing to the last minute and hoping it will 'be all right on the day'. We must be quite clear in our minds as to what action we intend to cover and how we plan to do it. The first move is to visit the church and spy out the land; note such points as where the sun will be (assuming it's shining!) at the critical time and select the best vantage points for the essential action — the arrival of the Bride, the Bridal Party entering the church, the happy couple emerging after the ceremony.

**2** A wedding is an occasion with a fairly formalised structure consisting of a series of events — preparations, church service and reception — which provide a natural beginning, middle and end. We should use this structure as the basic framework when planning our video coverage. An interesting opening sequence could show invitations being written and put into envelopes, a wedding dress fitting session or the Bride having her final hair-do. As such action, unlike events on the Great Day itself, can be directed to a certain extent, it's possible to introduce a little more artistic contrivance — such as asking for a piece of business to be repeated for matching a cut when editing (see Chapter 11).

**3** Be sure to arrive at the church in good time — at least three-quarters of an hour before the service is due to start (longer if lights need to be set up inside). This will allow you an opportunity to take up your chosen camera position to cover the arrival of guests before any of them actually appear so that you won't risk missing anybody. Use a tripod for this coverage and position it so that you can pan from the church gate to the church door itself. Not that you need to be constantly making such a camera movement (heaven forbid!), but such a set-up will allow you to take reverse angles of people walking up the church path which can be very useful for avoiding the jump-cuts that inevitably occur when the camera is stopped and restarted from the same angle. Apart from ensuring picture steadiness, having the camera firmly on a tripod will remove any temptation to 'cross the line' (see Chapter 7) and so inadvertently reverse the direction of the action on screen. The insertion of telephoto close-ups in such a sequence can also help to bridge awkward jump-cuts.

**4** The key moment of this pre-service sequence is, of course, the arrival of the Bride; be sure you're ready to get a good shot of her car approaching and pulling up at the church gate, then if necessary quickly change position to record her getting out of the car — a low-ish angle is usually advisable here because it will give the best view of her dress. Let her walk out of the picture as this will enable you to cut to her entering frame at the church door, so eliminating without a jump-cut the time-consuming and not terribly interesting walk up the church path.

The Groom will already have appeared on the scene (usually accompanied by the Best Man) and his arrival should also have been covered. Just a word of caution here: Best Men often have a somewhat heavy-handed line in humour and given half a chance will start hamming it up for the camera with exaggerated comic business. However, cheerful faces and natural smiles are generally much more effective, so have a discreet word with the BM beforehand if he's known to be an inveterate joker.

**5** If the couple wish the ceremony itself to be recorded, check that the officiating minister will allow this. If so, will there be sufficient light? If there isn't, will it be permitted to bring in lighting units and are there convenient mains points into which to plug these? Any extra lighting should be concentrated near the altar and should be of a diffused type so that no hard shadows are cast and nobody is dazzled.

The acoustics of churches are often far from ideal for sound recording because of reverberation from hard stone walls and floors, so it will be essential to get the mike in close to the couple and minister if the words of the service are to be clearly captured. Another possibility would be to use a highly directional gun mike from the camera position if this is reasonably near to where the action is taking place.

Should it prove impossible, for whatever reason, to shoot the ceremony, this very obvious gap in the proceedings can be effectively bridged by making a sound-only recording and later using this to back exterior shots of the church, stained glass windows and similar features that can, if necessary, be taken on a separate occasion.

**6** When taking shots of the Bride and Groom after they have emerged from the church, there is a tendency for inexperienced cameramen to overdo the use of the zoom lens. By all means use it once to zoom in on the happy couple when they first appear — this would be quite appropriate and effective. But don't use the facility to provide a 'trombone' obbligato to the Wedding March; it can be employed far more profitably to frame shots properly between takes.

**7** The advent of video notwithstanding, a feature of any wedding is the still photographer. His activities can prove a godsend to the movie maker by providing some appropriate action for cut-aways to link what might otherwise be a somewhat unsatisfactory series of artificially posed shots. He will be busily engaged arranging groups and making adjustments to his camera, so we can take interesting shots of him and intersperse these with the essential but rather static shots of the couple and the other principals in the 'drama' — the mothers and fathers, the bridesmaids, pages and so on. In this way we create a real movie sequence with a feeling of logical progression from shot to shot — and no jump-cuts!

**8** The photography session is a good opportunity for the video maker to get all those shots of relatives who will expect and be expected to feature in the movie. To make sure nobody who should be in is left out, it's as well to obtain a briefing in advance from whoever is commissioning the video. The justification, from a movie making point of view, for these personality close-ups is that the family members concerned are presumed to be looking at the Bride and Groom and admiring the former's wedding dress. To create this illusion, make sure you get the eyelines consistent — say, couple facing to the left and onlookers always looking right — by remembering the 180-Degree Rule as explained in Chapter 7.

**9** As we've seen, the photography session is, by its nature, bound to be a rather static sequence — something that's always a bit worrying in a movie. However, it can be given a lively conclusion by treating the inevitable confetti-throwing with imagination. This is where we can legitimately take the camera off its tripod to get in close to the action — in fact, a little controlled unsteadiness will enhance the spirit of the occasion in the video record. Don't worry if an occasional head bobs in front of the lens as this all helps the feeling of joyful confusion and gaiety when the shots appear on screen, as does the colourful 'snowstorm' of confetti as it swirls in diffused flurries in the foreground. Shots in this sequence should, of course, be brief to match the lively pace of the action.

**10** We naturally want to make certain that the concluding reception sequence is not an anticlimax. The best way of ensuring this is to concentrate on the key moments only — the cake-cutting, telegram-opening, highlights of main speeches and toasts — so that the sequence doesn't get bogged down in a lot of nondescript action which can quickly become boring. Once again, jump-cuts can be avoided when 'telescoping' events by using cut-away shots of guests (another useful opportunity for getting everyone in).

And try to finish off the movie with an exciting visual flourish. For example, you could line up a close shot of the rear of the going-away car with a 'Just Married' sign occupying the centre of the picture. As the vehicle moves slowly forward, zoom in at the same pace so that the sign's relative size is maintained. Then, as the car gathers speed, zoom back quickly — to give the impression of a 'jet-propelled' send-off to future happiness.

## . . . Making a Holiday Video

**1** The first and most basic requirement for making a successful holiday video is good picture and sound quality, and to ensure this you should check that your equipment is functioning properly. So a few weeks before you're due to set off — particularly if your camcorder has been lying idle for a while — make some test recordings to satisfy yourself that it's giving its optimum performance. If it's not, there will still be time to do something about it before you leave. Even if all seems to be as it should, it's wise to examine the lens and tape handling parts to make certain these are clean (see *Ten Tips on Camera Care*).

**2** Ever since there have been such things as home movies, those taken on holidays have generally had a 'bad press' from luckless friends and relations who've been obliged to view them. Remember the old joke? 'Care to come over and see my holiday movie?' 'No thanks — I've seen my own!' But why have they acquired such a bad name? Because an awful lot have been very boring, and I believe the main reason for this is that they lacked a proper *structure*. The viewers didn't feel they were getting anywhere because the movie had no real beginning, middle and end, so they quickly lost interest and just longed for the whole sorry affair to be over. The top priority, then, is to give our holiday video a clearly defined shape.

**3** But having agreed that we need a structure, how do we create one? There are two possible approaches. On the one hand, we can attempt an all-over view of the holiday by working around a particular theme, or on the other, concentrate on just one specific, but typical, episode and give this a detailed treatment. Depending on the nature of the holiday and the opportunities it offers, either approach can result in a highly rewarding movie record, as the following two tips will, I hope, show.

4 If opting for the 'theme' treatment, you should try to decide what this is to be before you leave home; such pre-planning will allow you to think about the subject matter and perhaps do a little research about the region you intend to visit which may help to spark off some ideas. Obviously, your theme will be governed to a large extent by the nature of the resort you intend to visit. It might have some famous historical associations that could give you a lead, or maybe there's some interesting local craft to suggest possibilities. If there's nothing of this sort ready-made, you could invent a theme of your own. Perhaps your holiday resort is one of those Spanish fishing villages that, in recent times, have become popular tourist centres with the character of the place being completely transformed. The old fishermen still ply their trade, but how do they feel about the changes? Visually, their traditional activities could be contrasted with today's tourist pursuits.

Another option is to make the theme a more personal one. Perhaps you are revisiting a location that you knew as a child. In this case the angle could be that you're rediscovering the place in the company of a companion to whom it is unfamiliar territory, with all the possibilities this opens up for suitable reminiscences on the soundtrack.

5 If we opt for the typical-single-episode approach, we shall need to think carefully about which occasion to choose; it needs to be one that, although only a portion of the holiday, is sufficiently representative to capture its flavour as a whole. Equally important, however, is that it has built-in movie potential, which means a reasonable amount of action and opportunities for securing interesting visuals. Such occasions might be the ascent of a mountain by rack railway and/or cable car (good chances for impressive travelling shots here), a trip on a lake or river steamer, or a visit to some site of great historical interest. But whatever it is, remember what I said in Chapter 1 and *think in terms of sequences* and not just individual shots that 'look nice' in themselves.

6 Because we decide to make a single-episode movie, that doesn't necessarily limit us to taking all the material on that one specific occasion. A little artistic licence is allowable here and all that really matters is that it *looks* as if the events depicted are linked in correct chronological order. For instance, you might decide to represent a seaside holiday with one comprehensive beach sequence, in which case you could use material shot on, perhaps, two or three different days — provided you had taken the elementary precaution of ensuring that your 'cast' were wearing the same clothes. And for real finesse it would also be necessary to see that all the scenes were taken under similar lighting conditions.

Referring back to the travelling sequences mentioned in tip No 5, a moment's thought will show that we can't be taking shots *of* and *from* the vehicle on the same occasion — hopping on and off could prove decidedly risky in most cases! On the other hand, such shots from these complementary viewpoints are essential if a fully-rounded impression of the journey is to be created. So here we are obviously forced to take the required shots out of sequence on different occasions and intercut them later — that is, assuming we have facilities for post-production editing as described in Chapter 11.

7 When shooting your holiday video always bear in mind that viewers, being fellow

members of the human race, tend to find faces more attention-grabbing than places. The simplest way to provide this essential element of 'human interest' is to include plenty of close-ups of family members reacting to the sights they are seeing when on various excursions, etc. People can also act as useful continuity links to move the action from one location to another. You could plan a transition sequence along these lines.

LS. *Family in a Florence square by a fountain.*
MS. *Mother takes a coin from her purse and tosses it out of frame towards fountain basin.*
CU. *Surface of water; coin plops in . . . then zoom back and tilt up to a long shot of Venice's Rialto Bridge with family standing prominently in the picture.*

**8** Although I started off by saying that we should plan our movies in advance as far as possible, we also need to be ready for the unexpected incident when it presents itself. I can quote a personal example. Some years ago I was shooting on the Fifteenth Century stone bridge at Wadebridge, in Cornwall, when a swan suddenly flew over the parapet and landed in the middle of the road, causing general confusion to the motor traffic. Luckily my camera was at the ready and I secured some quite good shots of the bird waddling with bland unconcern among the slowly moving cars and its confrontation with a workman who tried to shoo it back into the river.

The incident happened so quickly that I had to

*You just can't go wrong with shots of animals and children. Studies like these and the shot on the next page, for example, will add atmosphere and audience-grabbing interest to any holiday movie.*

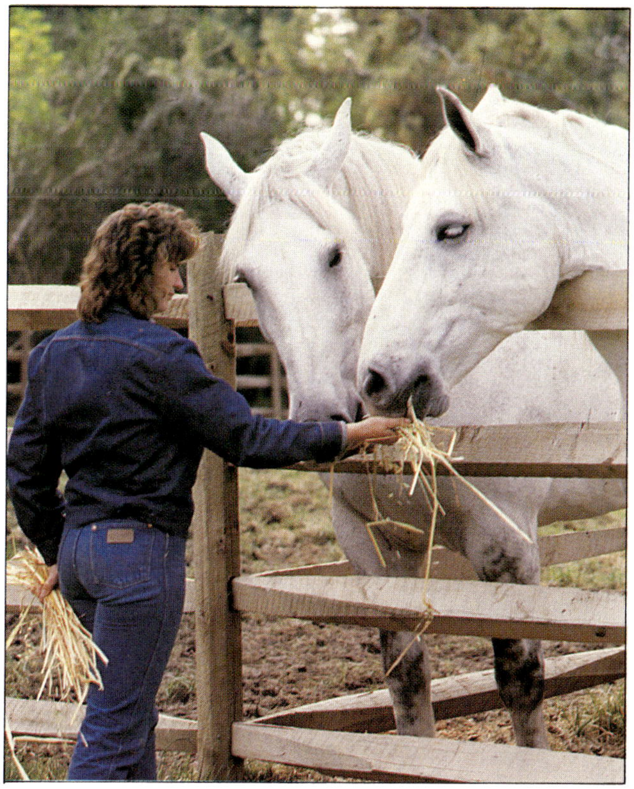

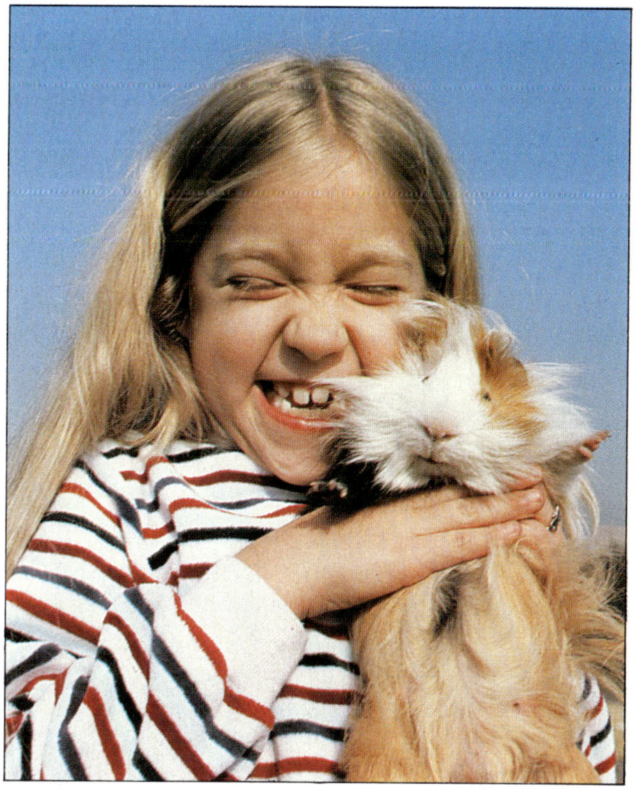

15

shoot first and think about how the shots would fit together afterwards. Needless to say, as taken, they were a string of awkward jump-cuts, but the answer was to intercut close-ups of my wife reacting to the swan's adventure. Of course, I didn't have these because while she *was* reacting, I was busy shooting the bird itself. So I persuaded her to repeat the appropriate expressions for the benefit of the camera, and later inserted these shots in between those of the swan. In this was I was able to turn a collection of discontinuous shots into a properly flowing sequence held together with those basic movie binding agents action-and-reaction and question-and-answer (see Chapter 1).

**9** As beach sequences are such popular subjects for holiday movies, let's consider for a moment how they can be handled for maximum impact. The secret lies in making all action seem *purposeful*. By this, I don't mean that it necessarily has to be frantic; if it can be shown that Father is dedicated to nothing more strenuous than getting a quiet sleep in his deck chair, that would be fine as a theme — and could be a good running gag if his objective is constantly thwarted by youngsters who pour water over his feet or bombard him with the beach ball.

Even in this TV age, kids still love the comic violence of the Punch and Judy show. The cry of 'That's the way to do it!' never fails to raise shrieks of delight as Mr Punch dispatches another victim with his big stick. The laughing faces of the young audience (or perhaps their sober rapt attention) make wonderful camera studies if intercut with the puppet action on stage — the Punch and Judy man is unlikely to object to your taking such shots if you put a little extra in the hat when it comes around.

Those with young children will want to devote a sequence of sand-castle building. When shooting this remember to keep in close to the action for most of the time so that facial

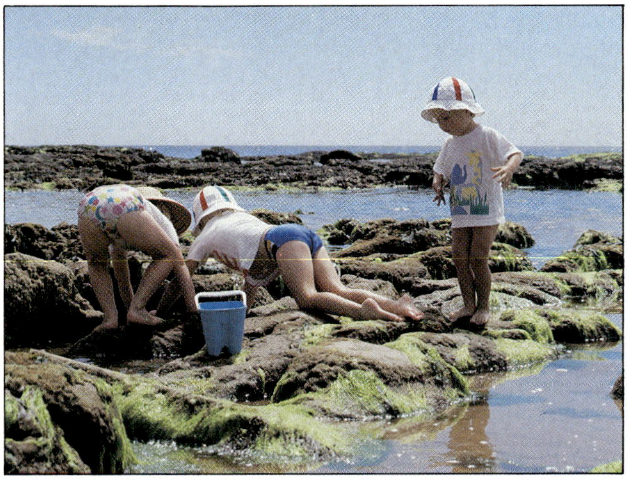

expressions can be clearly seen, and beware of jump-cuts. They can often be disguised by changing the image size and camera angle (see Chapter 8), but as I illustrated in tip No 8, they can be cured completely by using a cut-away between the two shots that cause the jump. Here, it might be a shot of Mother watching the castle-building with interest or Dad dozing in sublime indifference. Always try to end a sequence with a feeling of climax. In this case we might conclude with a close-up of a toy flag being stuck on top of the completed castle; then, as it flutters bravely, we could zoom back to show the edifice in all its glory.

**10** In my view, it's wise not to worry too much about the live sound when shooting a holiday movie; there may be the odd occasion when we need to have lip-sync — say, some vital words of explanation from a guide conducting us on a tour — but in the main, we shall get a much better soundtrack if we put on most of the sound after the video has been finally edited. The great advantage of doing it this way is that the sound — background effects, music, commentary or whatever — will flow uninterrupted over the edits to give a smooth and generally professional effect.

### . . . Recording a Sports Event

1 As with weddings, the secret of success with sports event videos is *pre-planning*. If it's at all possible, visit the track, course or stadium beforehand; try and visualise the action that will take place on the day and the best vantage points for covering it. For certain types of event, practice or time-trials are held and these will give you a chance to rehearse your camerawork and view the results. It's also a good idea, if practicable, to get a programme of events from the organisers so that you can make advance plans about what you'll need to shoot and when.

2 As often as not, your sports event will be taking place during daylight hours so there should be no problem about lighting. Where events are held after dark by floodlights, these are nowadays usually balanced to suit TV cameras so you have no worries — provided the white balance of your camera has been correctly set.

3 The professional broadcasting organisations achieve their slick coverage of sports functions with multi-camera set-ups. This is a luxury that's denied to most amateurs, so it's necessary to make good the deficiency by being particularly careful about the choice of camera positions. How you decide on these will be governed to a large extent by the kind of video record you hope to make. If the aim is a general impressionistic record that captures the atmosphere and sense of occasion, this can best be done by keeping the camera mobile and getting as much typical coverage of what's going on as possible. It perhaps goes without saying that with this approach the movie is really made at the editing stage and the facilities must be available for this to be done properly (see

15

Chapter 11).

Then again, if you wish to make a more or less complete record of specific events, one really advantageous viewpoint must be selected to cover each one adequately. For most events (high jumps and diving would be obvious exceptions) the higher the camera position, the better. This would allow you to begin the sequence with a wide-angle shot looking down on the track or arena to orientate the viewer. Then, as the action gets under way, you can 'move in closer' by using the telephoto setting of the camera's zoom lens.

4 For track or course events, a good camera position is on the inside of a curve, somewhere near its radial centre. This affords three very useful views of the action: oblique towards-camera shot of the subject as it approaches, a panning shot as it passes around the bend and a shot of it receding. Such a position gives you the option of covering the action in one continuous take or of breaking it down into three separate shots, perhaps intercut with reaction shots of spectators.

When shooting in this type of situation it's helpful to have an assistant further up the course who can signal the approach of the subject before it actually comes into your field of view. This will enable you to start the camcorder a couple of seconds early to allow for its backspace time (see Chapter 11) so that you don't miss the start of the action.

*When panning (see Chapter 6) with a moving vehicle or runner in some track event, always leave more space in the frame ahead of the subject than behind it. This avoids the unsatisfactory feeling that it's 'gaining on the camera' and about to dash out of the frame.*

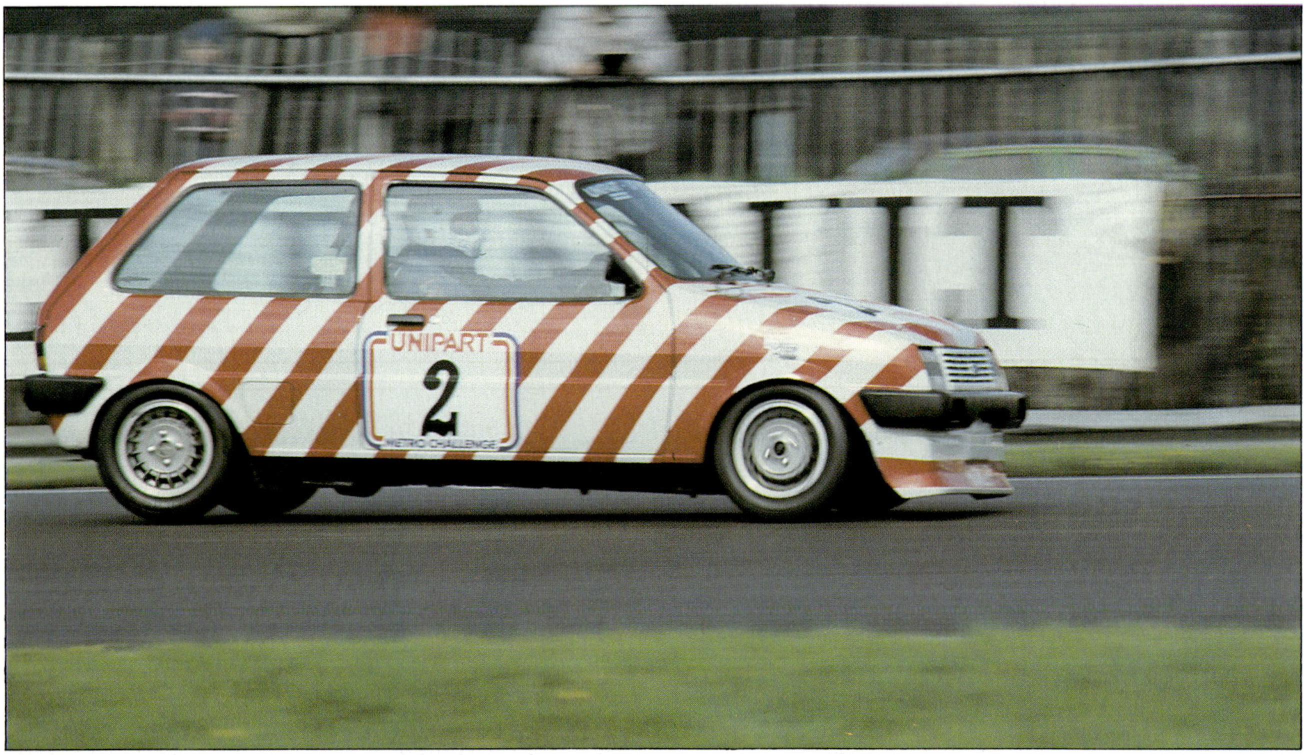

*When recording sports events, try to capture plenty of exciting action shots such as this. Here, the low angle chosen adds greatly to the dramatic impact, while the flurry of white show contrasts effectively with the rich blue sky.*

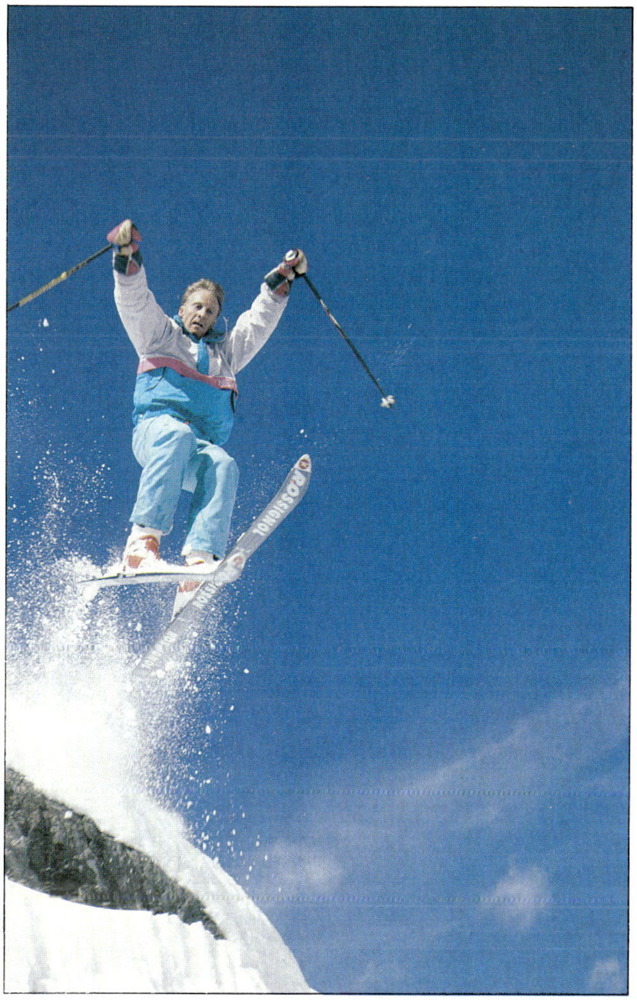

**5** An assistant will also be useful when it comes to recording effective 'live' sound. This is because best results will be obtained by having the microphone on an extension lead. Your helper should hold the mike — a unidirectional one is usually preferable —as close to the required sound source as possible.

For certain types of subject, general unsynchronised background sound for dubbing on to the picture after editing may be more suitable; this can be recorded quite easily on a portable audio cassette recorder which can be left running continuously for as long as is necessary.

**6** With events where it's not possible to get up close to the action, you will obviously be making a great deal of use of the telephoto end of your zoom lens. For such shots to be really satisfactory, they must be taken with the camera mounted firmly on a tripod, otherwise the inevitable camera-shake will seriously distract viewers from what they're supposed to be watching. Also remember (see Chapter 4) that the foreshortening effect of the telephoto setting will cause action towards or away from the camera to be apparently slowed down, so if an impression of speed is important, don't shoot head on.

**7** In tip No 4 I made a passing reference to reaction shots of spectators. These, as we have seen (Chapter 8) are an invaluable continuity aid as they can be used to bridge jump-cuts and condense sequences that would otherwise become boring because of repetitive or uninteresting action. So always take plenty as opportunity offers — you'll find they're absolutely indispensable at the editing stage. But when shooting them, remember the 180-Degree Rule (see Chapter 7) so that the spectators are facing in the right direction and really appear to be looking towards the event they're supposed to be watching. Providing this rule is observed, you can if necessary 'cheat' by cutting in shots of spectators who weren't present at that particular event at all!

**8** Subjects, such as track events, where a moving subject needs to be held in shot will call for panning (see Chapter 6). The object here must be to ensure that the camera movement is kept regular and smooth so that the subject remains in the same relative position on the screen throughout the shot. To avoid a jarring effect on the viewer, it's a good plan to start the panning movement just before the subject enters frame. And when composing the shot be sure to leave more space in front of the runner, racing car or whatever than behind it so that it doesn't look in danger of dashing out of the picture. For successful panning on the telephoto setting, a tripod is essential.

9 Winter sports make good movie subjects, but recording in the snow does pose certain problems for the video maker. It's in the nature of things that the location is likely to be cold, so make sure you're comfortable by being appropriately dressed. Thermal underwear is a good idea, as are wind-proof over-trousers and a quilted anorak; if the latter has a hood, make sure it's large enough to accommodate headphones for sound monitoring. You'll also need to keep your hands warm; gloves will make it difficult to operate the camera's controls so those specially slit mittens which allow fingers and thumbs to be brought into the open when necessary are the best bet.

The main technical problem of shooting amid wide expanses of snow is exposure — in fact, it's one of those very rare occasions when we may be in difficulties through there being *too much* light! This is because snow on the ground is such a brilliant reflector and causes the lens aperture to close right down, with the result that while the background snow is more or less 'correctly' exposed, the foreground subjects we want to record will be very much under-exposed, with murky colours and dark, unnatural complexions. The answer here, as we found in Chapter 2, is to use the backlight button or manual aperture control to adjust the exposure to suit the subject, even though this means that the background is then over-exposed.

I should stress that this technique is only appropriate for close or medium shots where detail in the main subject is visible; for really long shots, where the snowscape itself *is* the main subject, the camera's automatic system can be left to cope — although a neutral density filter (which cuts down the light entering the lens without affecting colour) is useful as it saves the lens having to work at an impossibly small aperture.

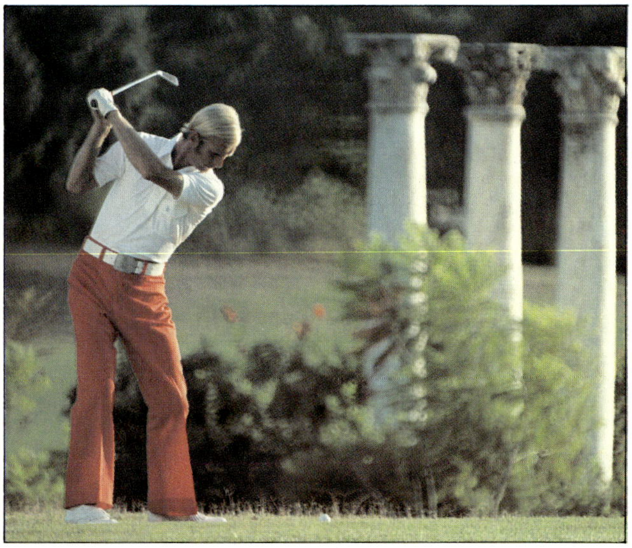

*As well as illustrating the theme of sporting videos, this shot of a golfer making a swing is a good example of a neat composition. The inexperienced camera user might have been tempted to put the figure centre frame, but a much more pleasing effect has been achieved by offsetting it to the left and balancing the composition with the background columns, right. Note, too, how the main centre of subject interest — the upper part of the golfer's body — occupies one of the four areas recommended by the Rule of Thirds (see chapter 9).*

10 If your equipment has the facilities, don't forget to add such professional touches to your sports video as superimposed captions and electronic stopwatch timings. These will make the record more interesting by telling your viewers exactly what is going on, and showing them the comparative timings achieved by competitors in those events where such things are of prime importance.

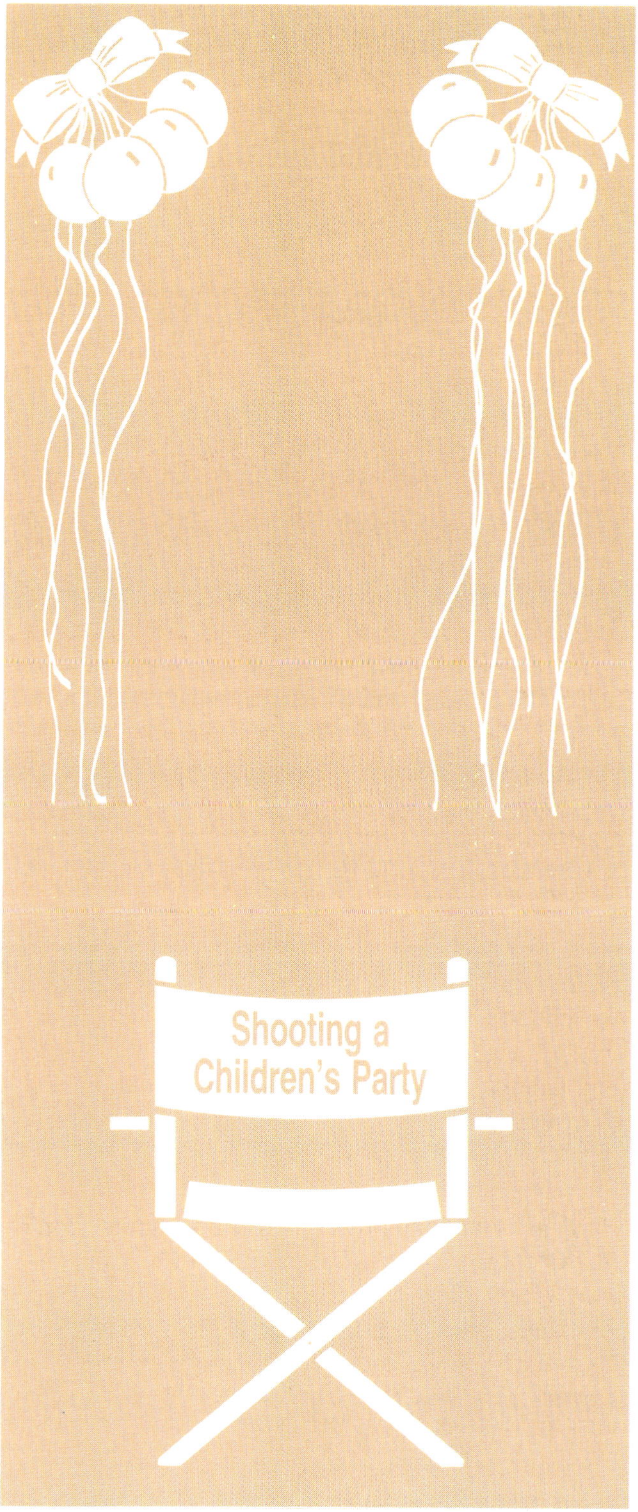


### . . . Shooting a Children's Party

**1** By its very nature, a children's party is bound to be a pretty boisterous affair. But, as movie makers, it's our job to try and bring order out of chaos, and one way we can do this is to make sure our video record has a firm shape — in other words, a proper beginning, middle and end. How we achieve this will, of course, vary according to circumstances, but possible ways of starting the movie might include a sequence of invitation cards being addressed and posted, Mother preparing the goodies for tea — or even Father getting ready to make himself scarce for the duration!

This could be followed by a short sequence of the young guests arriving, which would give a good opportunity for individual shots showing who is actually present at the party. It also allows the children to get over the novelty of seeing the camera in action before the fun and games really begin.

**2** You'll find life is made a lot easier on what can be a tough assignment if you have a callaborator. This should be a 'jolly uncle' (or auntie) type who is popular with the kids. He can be a kind of focal point for their attention and thus allow you to go about your business more or less unmolested. Not only this, but such a person can also be a useful continuity device when it comes to editing the movie; if there's an awkward break in the flow of the action — just cut to 'Uncle' to bridge the gap.

If 'Uncle' (as is sometimes the case) is also a bit of a sleight-of-hand merchant, he may be able to do interesting tricks with handkerchiefs, balloons, and so on, to entertain the youngsters. This will give you a chance to take some attractive candid close-ups of amused and amazed expressions.

# 15

3 With the low-light facility offered on many cameras, the video maker can shoot away quite happily by available light, confident that the picture will be adequately exposed. But as we've already seen, this useful feature is achieved at a price — inferior definition to that obtained with the 'normal' setting in brighter light. We shall also have to be very careful about focus, because there's very little depth of field when we're working at close range with the lens aperture opened up to its fullest extent.

If you have the opportunity, it would be a good idea to do a test in advance to see if the available light will give sufficiently good results; if it won't, then some supplementary lighting will be needed. For this type of situation where the action is constantly on the move, general bounce light (see Chapter 10) is probably the best solution — obviously with a lot of children milling about, it would be very inadvisable to have photographic lamps on stands with cables trailing around. Should the bounce lighting need an additional boost, you could have an assistant holding a battery operated video light for certain key moments — say, the cutting of a birthday cake.

4 I mentioned earlier that your movie should have a beginning, middle and end. Well, with a production of this nature most of the 'middle' will be taken up with party activities such as games. Children's games that involve running about are liable to get rather out of hand, and as it is obviously impossible for the cameraman to control the action, very little useful footage is likely to be obtained from such activity. It's therefore much more sensible to concentrate on games that are played sitting down, and the most rewarding of these from the movie maker's point of view is that old favourite Passing the Parcel.

Here the children will be sitting in a circle, and with the zoom set towards the telephoto end it's a simple matter to follow the progress of the parcel from hand to hand. You can also add variety and a touch of suspense by cutting in close-ups of the hands of the pianist (or controller of whatever apparatus is being used to provide the music). Then at the exciting moments when the music stops, you can concentrate on close-ups of frantic figures tearing at the parcel wrapping.

When the parcel at last yields up its secret, you can zoom into a close-up of the prize, following this with one of the delighted winner. It's action of this type — shown in big, bold close-ups — that really captures the spirit of the party.

5 If the party is a Christmas one being organised by some institution such as a nursery school or children's home, it's quite likely that the proceedings will be enlivened by a seasonal visit from Santa Claus. Such an event provides another focal point of interest around which the thoughtful movie maker can construct an effective sequence — plenty of opportunity here for more telling close-up studies of happy faces. If you get the chance, it's a good idea to have a few words with Santa before he makes his appearance, so that you can tip him off as to where to stand and which way to face for the best angles. As he makes his entrance, shoot him from a low angle; this will help suggest the child's eye view and make him appear even more awe-inspiring and impressive.

6 I've already mentioned that activities which involve running about are likely to result in confused footage. But if you really want to capture some fast action, then have your camera at the ready when tea is announced — but don't get carried away in the rush!

Compared with what you have had to cope with up to now, recording the meal itself should be comparatively simple. And this is probably the one occasion when it will be advantageous (and practicable) to use a tripod. This will help you to secure steady telephoto shots of the youngsters

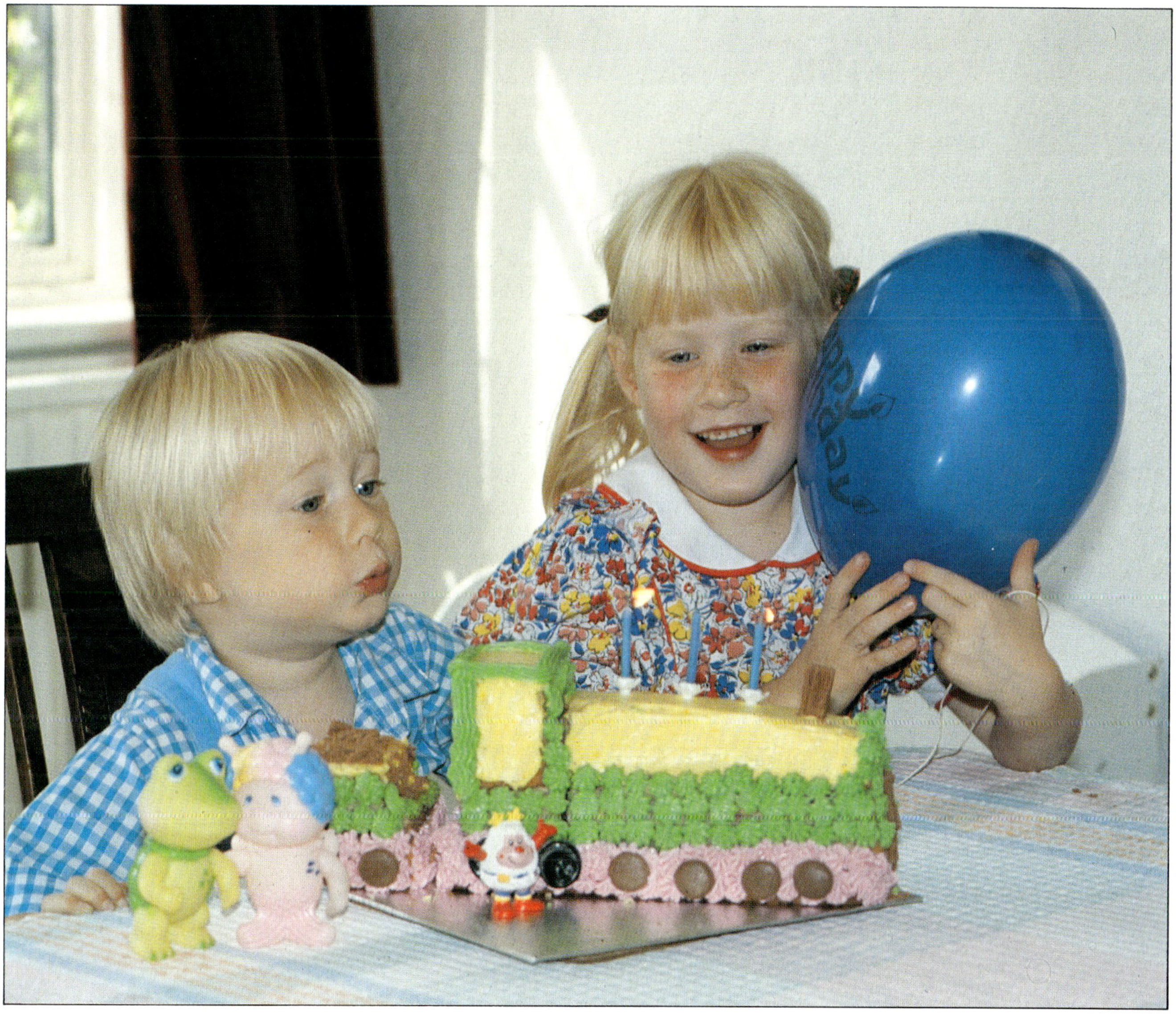

at their favourite pastime — cramming themselves with food. So once again concentrate on the bold close-up — cheeks bulging with cake, the quivering fragment of jelly on the tip of a nose. Also be on the lookout for those quaint and amusing bits of action that invariably occur when children are eating together.

If, as often happens on these occasions, the meal is rounded off with a session of cracker-pulling, this offers ideal material with which to end the sequence — and possibly the video itself if nothing more suitable crops up. To capture the feeling of excitement and surprise, the shots should be tightly framed and kept short.

**7** At most children's parties, once the young guests have had their tea, the host mother tries to get them to simmer down before they're packed of home or collected by their parents.

This is often done by providing some sort of entertainment — a conjurer, perhaps. Such a show can provide the kind of material we need to give our video a proper ending. The camera should be set up in a position that will allow it to get plenty of coverage of the performer and the audience (to secure reaction shots). The fact that all the shots are taken from more or less the same position will ensure that correct eyelines are maintained and the children will always be looking in the right direction relative to the conjurer or whatever.

If the show is, in fact, a conjuring performance, it might be fun to conclude the video with a bit of movie magic — if you get the chance to enlist the co-operation of the magician. He could be asked to make a magic pass with his wand, whereupon his audience immediately vanish (achieved by having the camera mounted on a tripod, stopping it while the audience file out and starting it again to record the now empty seats). Then you could round off the gag by having him bow to the camera and disappearing himself (achieved by similar means).

**8** I haven't said anything yet about sound. This can be a bit of a problem at parties on two counts. First, as we've already seen (Chapter 12), microphones lack the selectivity of the human ear so are unable to distinguish the sound we want from the general background hubbub. Of course, a directional mike can help here to a certain extent, but it won't overcome the second snag. This is our old enemy the aural 'jump-cut', caused each time we stop the camera when there's continuous identifiable sound — here, it's likely to be background party music. So for this reason it's usually best to reserve 'live' sync sound for those occasions when it's really essential — a cake-cutting speech, perhaps — and later to overlay the general goings-on with suitable jolly music (or genuine background sound recorded continuously on a cassette at the time) using the audio dub facility.

**9** If you are the lucky (?) parent who is hosting the party, you might like to finish on a personal family note by showing your own tired but exultant offspring being tucked up in bed. Here you have the chance to use sound with a touch of imagination if you add a track of party tumult (recorded on tape at any time during the proceedings) to a shot of the child's head on the pillow. This would indicate that he/she was reliving the excitement of the day's events before finally falling asleep — which could be suggested by fading up a lullaby tune just before the final fade-out.

**10** And one final point which is, perhaps, almost too obvious to mention — but better safe than sorry! Make sure you keep all equipment in a safe place away from inquisitive little hands when it is not in use. This applies not only to expensive and easily damaged items, but also to videolights which can get very hot when they have been switched on for some time.

### . . . On Camera Care

1 Like any other piece of apparatus, a video camcorder needs the regular equivalent of a 'wash and brush up'. The cassette chamber can easily be cleaned with a soft brush — the type of blower-brush sold as an accessory by photo shops is ideal. It's important to keep this chamber clean because dust and particles of grit can cause trouble if ground between the tape and delicate recording heads. Any deposits that have built up on tape handling parts can be removed with the aid of a matchstick sharpened to a chisel shape. But never, repeat *never,* use any sharp metal instrument for this purpose.

2 Lenses need to be kept clean and blemish-free if their performance is not to be impaired. They should only be cleaned with proper lens-cleaning tissues and not any coarse material that could possibly damage the coating. Before applying the tissue to the lens, however, make sure that any loose particles of grit have been removed with a soft brush, otherwise there is a danger of the 'sandpaper effect' which can do serious damage to the delicate surface. You need to be especially wary if droplets of sea spray have caused small deposits of salt to be left behind. The crystals should first be dissolved by dabbing them gently with a moistened tissue *before* any wiping action is applied. Always wipe carefully from the centre of the lens to its edge.

3 Video cameras tend to be used most at holiday time, which means that they get involved in a certain amount of travelling. This doesn't pose any serious problems provided a few simple and fairly obvious precautions are taken. Although video equipment is reasonably robust, it's not a good idea to shake it about unnecessarily. Luggage racks of motor coaches

15

and the boots of cars are notorious 'rattle traps', so don't dump your camera into these unless it's well wrapped up to insulate it from the inevitable vibration. It's not usually advisable to pack your camera with your other luggage (baggage tends to be thrown around by porters and loaders), but if there's no alternative, make sure you place it between other articles with good shock absorbing properties, such as thick jumpers, and not touching the top, bottom or sides of the case.

4 Having arrived at your holiday destination — at home or abroad — you're naturally going to take your camera on to the beach. I don't wish to sound alarmist, but it needs to be said that this is the most hazardous environment it's ever likely to find itself in. Sand and sea are the two arch enemies of video equipment, the former because of its abrasive properties and the second because salt water is a corrosive agent. An effective way of protecting your vulnerable lens from both these perils is to keep a UV filter permanently fitted when using the camera at the seaside (such filters are normally used by photographers to cut down the haze caused by ultra-violet light when shooting over large expanses of open water or snow).

If you have to change a tape on the beach, make sure sand doesn't blow into the cassette chamber. The body of the camera can be protected from sand and spray by putting it in a plastic bag when not in use. If a couple of holes are cut for the lens and viewfinder eyepiece to protrude through, the camera can even be used without removing it from the bag if conditions seem particularly hazardous.

5 This is the tip that always raises eyebrows and often a laugh: if you have the misfortune to drop your camera in the sea then, at the very first opportunity, *plunge it into a bucket of fresh water*. Yes — I'm being quite serious. The point is that sea water, as already mentioned, is highly corrosive, so the sensible thing is to wash it out of the works as quickly as possible so that the salt has the least chance to do lasting damage. After this treatment you should get the camera into a service department as soon as you can.

6 Before leaving the subject of sad accidents, I recommend that the camera should always be carried in its case when not actually in use. This is a particularly wise precaution when proceeding over rocks or rough ground. It's a natural reflex action to shoot out your hand to save yourself if you stumble — and if that hand happens to be holding the camera, there could be some very unhealthy sounds as it meets a boulder! If your camera case has a shoulder strap, be sure to wear it diagonally across the body. Worn like this, the case is under much more control, and can't swing away from your body when you lean forward as it could if slung from the neck or shoulder.

7 One likes to think that fellow holiday-makers on the beach are as honest as oneself. But just in case some aren't, it's as well not to put temptation in their way by leaving valuable equipment on view and unattended. If you don't have anyone who can stand guard while you go for a dip, then at least make sure your camera is away out of sight in a carrier or under a pile of clothes.

8 While *you* may have gone on holiday to soak up as much sunshine as possible, don't make the mistake of thinking that it's a treat for your camera, too. Excessive heat is the very last thing that video equipment needs, so don't leave the camera lying exposed to the sun for long periods (or even short ones, come to that). The dark body casing will absorb the solar rays, turning the

tape chamber into a small oven. Tape is certainly not better for being baked!

And a special word of warning to those whose video cameras have a tube image sensor, as opposed to the CCD type. Like the human eye, such a camera should never be pointed directly at the sun and for much the same reason: permanent damage can be done to its 'retina'.

**9** Video equipment is not exactly cheap, so before proceeding on a holiday it's wise to make certain that it's adequately insured. Possibly you already have it covered under the 'special valuables' section of a general household policy, but if so, does this cover you for going abroad? Best make sure before you go.

And speaking of going abroad, it's sensible to take the receipt or some other documentary evidence to show that the camera was, in fact, bought before you left home. This can save possible problems with the customs officials on your return journey, as you will then have conclusive proof that you didn't buy the camera in a foreign land.

**10** After returning from any movie making expedition, if you think that you're unlikely to be using the camera again for some time, it's a good plan to remove the batteries before putting it away. There's no danger then that they could leak and cause silent but expensive damage while the camera lies undisturbed in the bottom drawer.

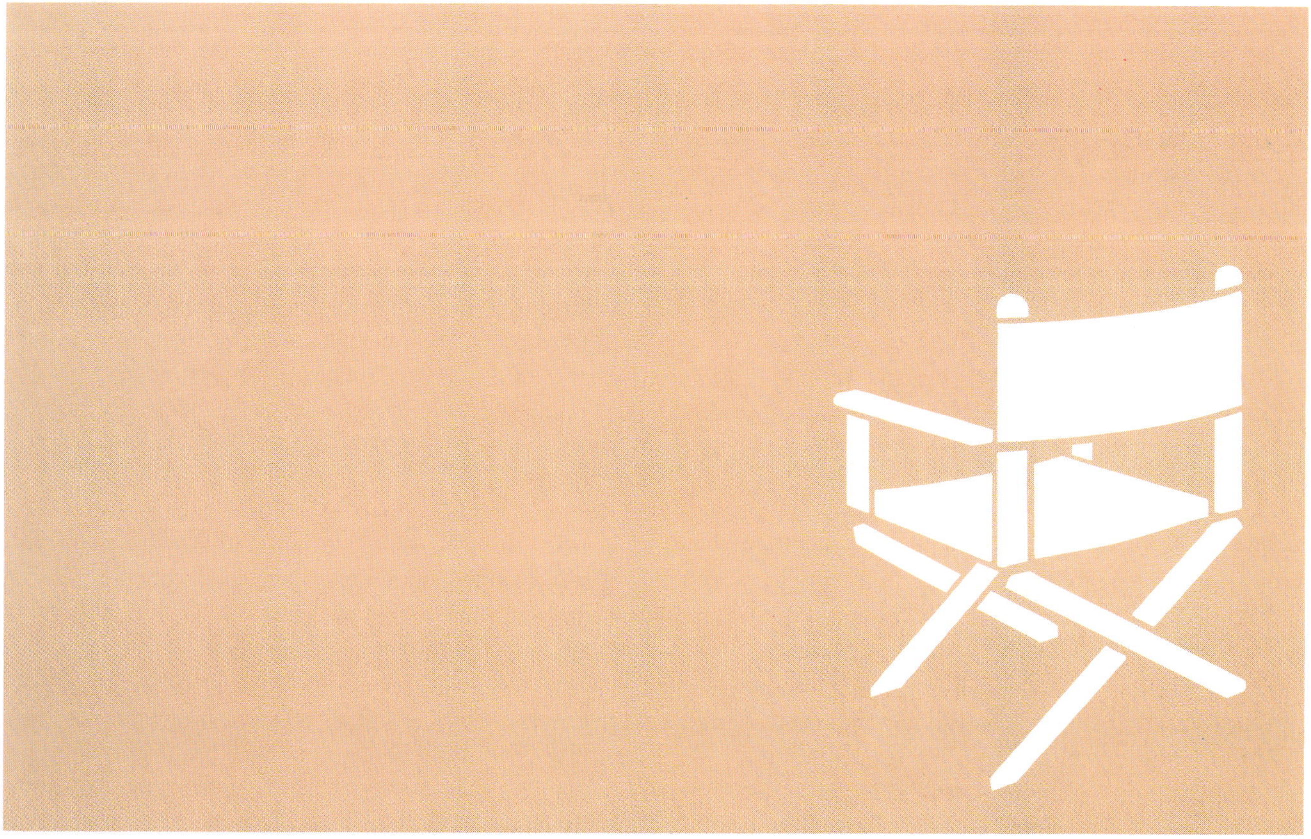

# 15

## . . . On Tape Care

**1** Video tape is a fragile and vulnerable material, so to protect your precious recordings from the ravages of time and other hazards, certain precautions need to be taken. Fortunately, these are simple and easy to remember.

Polyester recording tape has a tendency to shrink with the passing years — a process that's encouraged and accelerated by heat. So never store tapes in unduly high temperatures and keep them well away from radiators, fires and hot-air ducts. To be on the safe side, they should be kept in an environment that doesn't go much above 70 degrees Fahrenheit (average room temperature). Excessive heat can also have a bad effect on the plastic cassette by causing warping, which may then prevent the tape from running freely and winding properly with all the trouble that this implies. So as well as keeping it away from the obvious heat hazards mentioned above, be sure not to leave a cassette in the boot or on the seat of a car in summer when the internal temperatures can rise to oven-like proportions.

**2** If constant exposure to heat is bad for tapes, so also are rapid changes of temperature, as when a cassette is taken from a very warm to a very cold environment. In extreme cases this could cause cracking of the tape, so allow your recordings to acclimatise themselves literally by degrees.

**3** Still on the subject of environmental conditions, moisture is as much a deadly foe to tapes as extremes of temperature. It can attack in two ways: stealthily over a period when a tape is stored in an atmosphere where the relative humidity is high, or suddenly in the form of condensation when a cold tape is bundled into an already warm recorder. Lengthy storage in conditions of high humidity could cause the magnetic layer of a tape to become unbonded from the base and peel off, so it's recommended that cassettes are kept in an environment where the humidity doesn't exceed about 65 per cent.

If moisture forms on a tape through condensation when placed in the recorder, this can cause it to stick to the head drum which is obviously not good for either the heads or the tape. So when a tape is moved from a cold to a warm room, give it time to 'thaw out' and adjust to the new temperature before inserting it in the recorder.

Perhaps it hardly needs saying that tapes should always be kept well away from liquid hazards of all kinds, so for this reason it's as well not to park them on sideboards or tables where any drinks — tea, coffee or even more potent beverages — are likely to be served.

4 Dirt is also high on the list of tape perils. It can prevent proper tape-to-head contact, so causing drop-outs, and also do a lot of no-good to the delicate recorder heads themselves because a tape that carries a coating of dust and grit acts as a kind of sandpaper. Dust is always present in the air so cassettes should be kept in proper plastic cases with the tape aperture of the cassette away from the opening side of the case. Remember that the lower down any object is stored, the more dust it is likely to attract, so never keep your tapes on a low shelf — or, worse still, on the floor. And always make sure tapes are wound to the beginning so that any dust that might find its way into the cassette will be deposited on the leader rather than on the length that really matters. Dirt, in the shape of oxide dust and fluff, can also be picked up from the VCR itself; however, the risk of this can be minimised if the machine is regularly serviced and cleaned.

5 There are few things more impermanent than a magnetic recording. The principle that makes it so easy to put on in the first place, makes it equally quick and simple to remove — whether by accident or design. So it's very important that your precious tape is not brought into close proximity with a strong magnetic field which could partially or totally erase the recorded signal. Actually, it does take a fairly strong field to do this, but beware of placing tapes close to loudspeaker magnets or electric motors and transformers. However, I believe the alleged risks of carrying magnetic recordings on underground trains to be something of a myth as I've done this many times myself without ill effect.

6 A tape is rarely likely to be in more physical danger than when it's actually playing on the recorder, so it's important that any risk is kept to a minimum by ensuring that the machine is always functioning properly — regular servicing is the answer here. One possible source of trouble could be that the tape-tensioning devices have lost their correct adjustment, so causing the tape to wind in a slack, uneven way. This could lead to physical damage when it's played next time. Occasionally, slackness can be introduced into the tape coil if the VCR is stopped and restarted a number of times at short intervals. If you find that a tape does seem to have taken up in a sloppy fashion, the remedy is to run it fast forward until the end is reached and then rewind again to the beginning.

7 The pause button on a VCR is certainly a handy device, but it does impose extra wear and tear on the tape every time it's used. When you press it, the tape is stopped but the head drum continues to whizz around, so acting as a kind of 'grindstone' on the particular section of tape with which it's in contact. For this reason it's important that the recorder is not kept in the pause mode for longer than is really necessary,

and that the tape is not repeatedly stopped at the same point. For rather the same sort of reason, it's as well not to load and unload a tape at the same point several times, as this can cause damage to its edges which in turn will result in an impaired performance on playback.

**8** It's essential that the tape should always run freely, and the design of the cassette is a precision job to ensure that this happens. But like all such mechanisms, the cassette is sensitive to rough handling and should be treated with respect, so avoid subjecting it to excessive vibration or a drop from a great height. And it's best to store cassettes in a vertical position as this will reduce the risk of warping.

**9** Fingers off! — that should be the rule as far as tape handling is concerned. In fact, there's no reason why the user should ever need to touch the actual tape in the cassette at all, but just in case you're tempted, remember that grease from fingertips can be transferred to the magnetic coating and thence to the recorder heads which tend to be sensitive to such substances. So handle the tape by the cassette only.

**10** And a final obvious tip (or is it?). To guard against accidental erasure of an irreplaceable recording, make sure the 'safety' tab on the cassette has been removed.

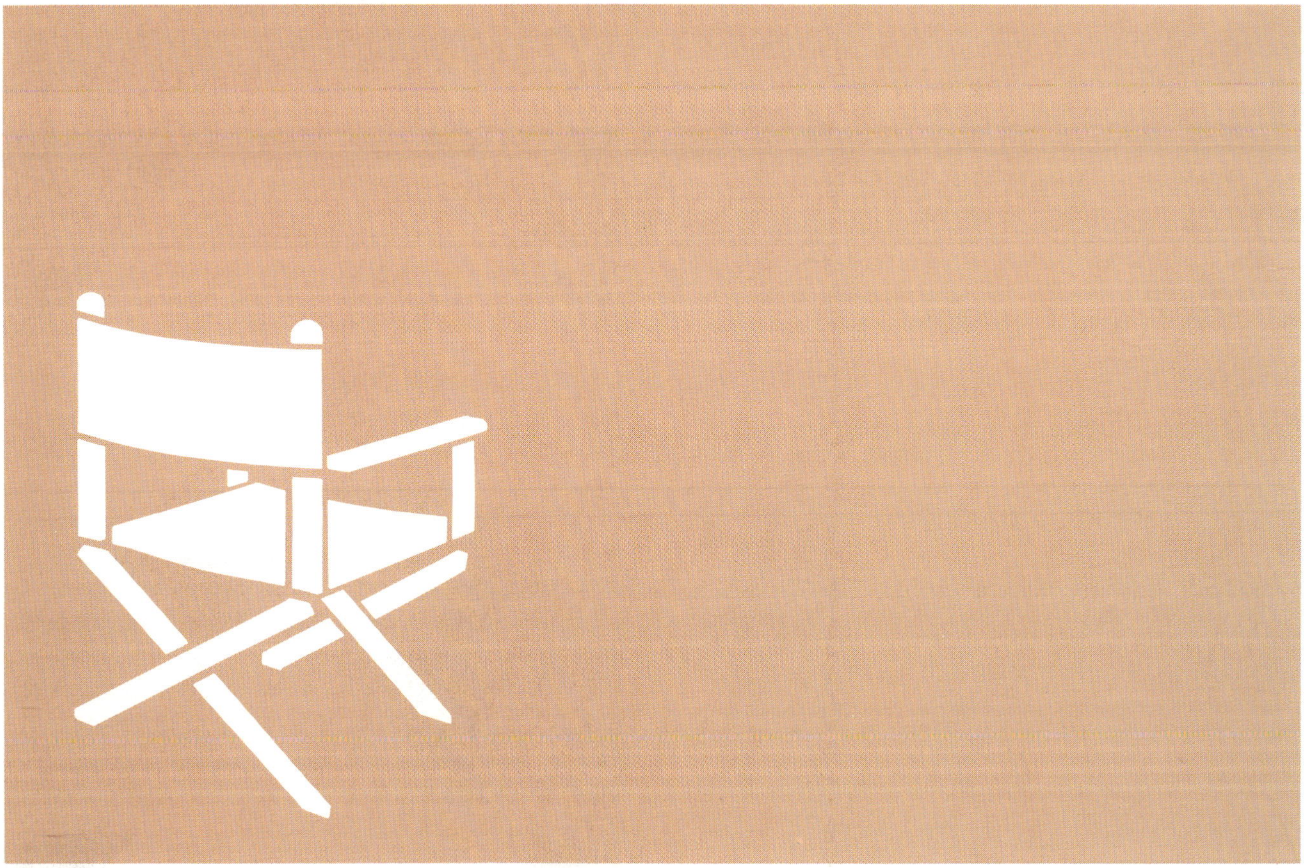

# Index